Not Here, Not Now

Not Here, Not Now

Speculative Thought, Impossibility, and the Design Imagination

Anthony Dunne & Fiona Raby

The MIT Press Cambridge, Massachusetts London, England

The MIT Press
Massachusetts Institute of Technology
77 Massachusetts Avenue, Cambridge, MA 02139
mitpress.mit.edu

The MIT Press would like to thank the anonymous peer reviewers who provided comments on drafts of this book. The generous work of academic experts is essential for establishing the authority and quality of our publications. We acknowledge with gratitude the contributions of these otherwise uncredited readers.

This book was set in KW Museum by Kellenberger-White.
Printed and bound in the United States of America.

Library of Congress Cataloging-in-Publication Data
Names: Dunne, Anthony, author. | Raby, Fiona, author.
Title: Not here, not now : speculative thought, impossibility, and the design imagination / Anthony Dunne and Fiona Raby.
Description: Cambridge, Massachusetts : The MIT Press, [2025] | Includes bibliographical references and index.
Identifiers: LCCN 2024024740 (print) | LCCN 2024024741 (ebook) | ISBN 978-0-262-04966-5 (hardcover) | ISBN 978-0-262-38238-0 (epub) | ISBN 978-0-262-38239-7 (pdf)
Subjects: LCSH: Design—Philosophy.
Classification: LCC NK1505 .D863 2025 (print) | LCC NK1505 (ebook) | DDC 744.01—dc23/eng/20241214
LC record available at https://lccn.loc.gov/2024024740
LC ebook record available at https://lccn.loc.gov/2024024741

10 9 8 7 6 5 4 3 2 1

EU product safety and compliance information contact is:
mitp-eu-gpsr@mit.edu

Contents

Preface

The ideas in this book took shape during a time when, for us at least, what we thought of as "reality," something stable, shared, and durable, began to feel like it was coming apart at the seams. So many things once thought impossible, unrealistic, or unlikely were becoming not only possible but real too. Surely, if these "impossibilities" were now "possible," then other more desirable kinds of impossibility might also be possible? We began to wonder what it meant to design at a time when, for many people, reality itself was becoming unrealistic.[1] And specifically, what this meant for the design imagination, where traditionally, clear borders keep the unreal, impossible, and unlikely at bay.

As designers fascinated by fiction, imagination, and different kinds of speculative thought, we were curious to explore this seemingly unstable boundary between the possible and the impossible a little further. Not to develop a new theory or approach, but a more nuanced understanding of the interplay between fiction and reality, ideas and things, the possible and the impossible, that collectively shape the design imagination.

We embarked on a journey that took us into a world of philosophical reflections on the nature of the real. The edges of science, where alternative and often competing models of reality flourish. The imaginings of writers who spend their days constructing expansive fictional realities. And the history of ideas that traces how what is considered real at any moment in time can fade in and out of existence, leaving behind infrastructure, institutions, ideas, and objects belonging to outmoded reality systems that shape and limit the reality they continue to exist within.

In a way, this book is a kind of conceptual travelogue, bringing together words, images, and objects that capture in design form some of the ideas encountered along the way, and our interactions with them. Ultimately, it is about imagination, specifically the design imagination, and loosening the bonds that tie it to a reality currently being contested on multiple fronts. A reality that for many people is broken, along with its accompanying forms of design. Preparing our minds for yet unknowable and even unimaginable new realities.

Although we hope this book will be of value to people working in a range of fields, not just design, it is specifically aimed at those, like us, grappling with design in metaphysically turbulent times, which it seems are set to continue for some time yet. As practitioners based in academia, we are always faced with the question of how to write. And the structure of this book is itself a design project that explores different ways of bringing ideas about impossibility into conversation with objects—actual and imagined, existing and invented, made from words, images, and matter—through notional archives, libraries, glossaries, taxonomies, lists, tales, and of course essays.

We hope readers will find it inspiring and generative, the beginning of a journey rather than an end. One that sparks many more investigations and explorations that expand the possibilities that arise when speculative forms of thought, from many different fields, meet design practices concerned with the design of things, objects—the stuff of everyday life.

New York, January 2024

Acknowledgments

Straddling the worlds of academia and design, we are fortunate to be able to develop our thinking and practice in dialogue with both communities. Although often contradictory, this tension significantly enriches our efforts to bring theory and practice into conversation through the design of everyday objects. We are appreciative too of having an academic home at Parsons, The New School, where we can combine the different strands of our thinking through teaching, research, and creative practice.

The ideas in this book have benefited greatly from conversations with many of our colleagues at The New School. In particular, Victoria Hattam, Joseph Lemelin, Dominic Pettman, Hugh Raffles, Janet Roitman, and Miriam Ticktin, who all generously engaged with our thinking during the formation of the ideas and projects in this book. We are also grateful to the many students who signed up for our classes, joining us as we grappled with the challenges of designing for the “not here, not now.”

On many of the projects in his book, we worked with research assistants who brought an invaluable creative energy to the research. We would especially like to thank Franco Chen, Lila Feldman, Del Hoyle, Joonas Kyöstilä, Devon Reina, Philipp Schmidt, and Jonas Voigt. Also, designers Sarah Hawes, who developed and made the communo-nuclearist garment, and Carolyn Kirschner, for her computer modeling and rendering work.

We are also grateful for the many opportunities to test and explore our thinking through seminars, workshops, and visits, especially at CERN, the Royal Melbourne Institute of Technology, and the Porto Design Biennale. In 2022, we were granted a sabbatical by The New School, which allowed us to carry out a significant part of the writing and some project work during a residency at the International Programme for Visual and Applied Arts in Stockholm, where we again benefited from rich and stimulating conversations with our fellow residents as well as faculty and students in the Master’s Programme in Design Ecologies at Konstfack, in particular with the head of the program Martin Ávila.

With the right interlocutor, interviews can be an effective way of uncovering and shaping half-formed ideas, and we’d like to thank James Auger, Julian Bleecker, Kathryn Johnson, Christopher Marcinkoski, Mariana Pestana, and Elvia Wilk for their insightful questioning and teasing out of ideas in progress during the development of this project.

Being concerned with materializing ideas in the form of physical objects, we are dependent on external sources of funding to bring our work to fruition. There is many a thank-you to be given here. In particular, Jan Boelen at Z33 in Hasselt, Belgium, and A/D/O in Brooklyn; Thomas Geisler at the Museum of Applied Arts in Vienna; Zoe Ryan and the 2nd Istanbul Design Biennial; the Mellon Foundation for a Sawyer Seminar grant; Katia Baudin at the Kunstmuseen Krefeld, Germany; Ewan McEoin and Simone LeAmon at the National Gallery of Victoria, Melbourne; Beatrice Leanza and the Museum of Contemporary Design and Applied Arts, Lausanne, Switzerland; Tim Marshall at the Royal Melbourne Institute of Technology; and Justin McGuirk at the Design Museum in London. All of them generously supported the realization of our work and provided opportunities to engage with a rich variety of audiences.

Seeds of some of the ideas in this book were explored in several shorter essays and texts that provided an invaluable opportunity to test our thinking while notions were still forming. A few short extracts from these appear in this book, including “The School of Constructed Realities” (Maharam, 2014); “A Larger Reality,” in *Fitness for What Purpose*, ed. Mary V. Mullin and Christopher Frayling (London: Eyewear Publishing, 2018); “Design for the Unreal World,” in *Studio Time: Future Thinking in Art and Design*, ed. Jan Boelen, Ils Huygens, and Heini Lehtinen (London: Black Dog, 2020); “Designing Fiction When Reality Itself Is No Longer Realistic,” a section of the essay “Material Imaginaries,” in *Perspecta 54: Atopia 54*, ed. Melinda Agron, Timon Covelli, Alexis Kandel, and David Langdon (Cambridge, MA: MIT Press, 2022); “Treading Lightly in a World of Many Worlds,” in *NGV Triennial 2023* (Melbourne: National Gallery of Victoria, 2023).

And finally, a big thank you to everyone involved in the realization of the book itself, especially the team at the MIT Press for taking on this project, and providing support and guidance. In particular, we’re grateful to Deborah Cantor-Adams, Noah Springer, and Matthew Valades; designers Kellenberger-White for their thoughtful and sensitive design; and Lisa Sneijder for her tireless work tracking down permissions.

1 Introdu
Realists
Larger Re

ction
of a
ality

Chapter 1

Introduction: Realists of a Larger Reality

Writing about design at a time when in the West, we are experiencing a major reconfiguring of geopolitical power relations and a shift toward inward-looking nationalism; accelerating marketization of seemingly every aspect of life, including education; increasing environmental instability and uncertainty; the realization that our techno-utopia might in fact be a dystopia; and the appearance of significant cracks in a political and economic system that while generating vast wealth, has completely failed to ensure its fair distribution, it is hard to be optimistic. But we will try.

It is hard also to focus on ideas that do not have immediate or obvious practical applications in the face of these challenges, many of which lie beyond the scope of design as a field. And although significant efforts are being made in areas where design can have an impact, such as rethinking the materials the human-made world is made from and their impact on the community of life we share this planet with, it is important to address the worlds we carry around inside us too. Worlds made from ideas, values, and belief systems from which new possibilities may eventually emerge, if sufficiently nourished.

Rather than more futures or end points extrapolated from a faulty present, new starting points are needed. Spaces to momentarily step out of existing realities, a “not here, not now,” to imagine different ways of being in the world, made tangible through the design of everyday things. Not as an escape, not as a vision of how things will or should be, but to shake off old habits, patterns, and mindsets in preparation for as yet unknown realities. As Timothy Morton so beautifully puts it, “There are thoughts we can anticipate, glimpsed in the distance along existing thought pathways. This is a future that is simply the present, stretched out further. There is not-yet-thought that never arrives—yet here we are thinking it in the paradoxical flicker of this very sentence. If we want thought different from the present—if we want to change the present—then thought must be aware of this kind of future. It is not a future into which we can progress.”[1]

As designers, when we design for the “not here, not now,” our work is by default relocated to a “future.” Particularly if technology is involved. But futures are just one way of framing the not here and not now.[2] And not always the most helpful for the kind of work we do. In contrast to futurology, future studies, and foresight, which all attempt to mitigate risk in the face of uncertainty or identify preferred futures, the future's primary value for us has been as a place for thought experiments that explore alternative values and worldviews made concrete through the design of everyday things. Not as an end point, but a starting point.

Fully colonized by the technological dreams of the entertainment and technology industries, the future as a concept for facilitating imaginative thought, in design, has become too restrictive, at least for us. Even its forms of representation have ceased to evolve, long ago crystallized into a hackneyed aesthetic by a conservative Hollywood production machine exemplifying what the late Mark Fisher has called the "monopolization of possible realities."[3] Futures no longer provide enough critical distance from the present even when set in a far future that blurs into fantasy.

Worlds within Worlds

On Friday, June 24, 2016, sixty-seven million people woke up in one of two United Kingdoms.[4] Each with its own imagined history and future. Overnight, the United Kingdom effectively split into two worlds. But what does it mean to talk about multiple "worlds" in this sense? Clearly it does not mean physically separate worlds, nor the infinite parallel worlds sheared off from this one, each time a measurement is taken, at least according to the Many Worlds Interpretation (MWI) of quantum mechanics. Nor do we mean the possible worlds of modal realism in philosophy, where for every possibility there is a world where it exists, just because we can think it. It is different too from the complex border that winds through the individual homes of Baarle-Nassau in the Netherlands and Baarle-Hertog in Belgium, where each house displays either a Dutch or Belgian flag next to its street number indicating which country the front door is located in. Perhaps it is closer to the surreal fusion of worlds in China Miéville's *City and the City*, a detective story set in a world where two city-states coexist, superimposed in one geographic zone.[5] Taught from early childhood to "unsee" the other world, the inhabitants of each are unable to see the other as they move in and out of shared buildings. Seeing anything belonging to the other world, or "breaching" as it is known, is the most serious of crimes. For us, a "world" somehow encapsulates a set of beliefs, ideas, hopes, and fears that provide a framework for making sense of things, of how things work, conceptually, as well as what is and is not possible. A sort of useful fiction linking individual minds to a larger collective imagination that shapes the contours of the possible.

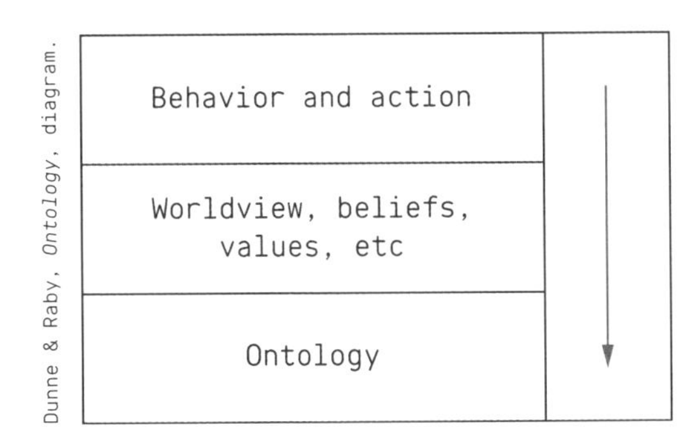

Dunne & Raby, *Ontology*, diagram.

If besides nourishing the worlds people carry around inside them, the purpose of designing for worlds that do not exist, in the here and now, is to prompt thinking about alternative ways of being in the present world and provide new vantage points that allow for a questioning of assumptions and values, then working with ontology as a conceptual raw material might offer more potential than futures.[6] Working at this level is unusual for design, which typically operates at the level of behavior, thought, and action. Shaping, aiding, guiding, nudging, and changing. At times it might touch on the level beneath this, worldviews, which inform our thoughts and actions. Beneath this is another layer, the place where ontologies are found, a place design rarely visits.[7] Home to fundamental shared assumptions about what can and cannot exist, and ways of being in the world. What is possible or not. To do this, we will need to explore ideas lying beyond design in fields such as literature, philosophy, anthropology, and the edges of science.

Shades of Real

A thinker we have found particularly inspiring for this project is late nineteenth, early twentieth-century Austrian psychologist and philosopher Alexius Meinong and his "theory of objects," which opened a door for us into a new conceptual world.[8]

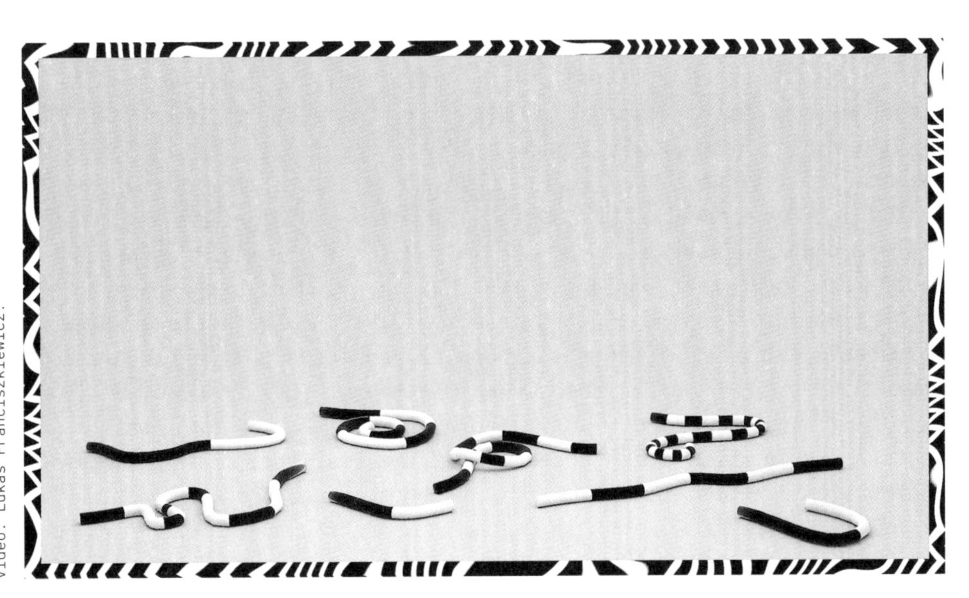

Dunne & Raby, *Meinong's Jungle*, video still, 2015. Video: Lukas Franciszkiewicz.

We first encountered Meinong's ideas while working on a project with the Museum of Applied Arts (MAK) in Vienna in 2014. We were one of ten designers invited by the MAK to select a collection of designs for *Exemplary*, an exhibition celebrating 150 years of the MAK and aiming to raise questions about the kind of objects a museum of applied art might collect over the next 150 years. Our collection consisted of fictional technology-related products and services developed by writers of social fiction over the previous 150 years, including passages describing credit cards and a twenty-four-hour live music distribution system from Edward Bellamy's *Looking Backward 2000-1887* (1888). In the exhibition, the books were presented with inserts framing passages describing the products. After the exhibition, the curator proposed adding our exhibit to the museum's permanent collection and had to present it to a selection panel. The panel could not accept that a museum of applied art should consider fictional objects made from words as part of its collection, even if they impact how people think about their daily lives and material culture.[9] In the discussion, someone made a connection to Meinong and his theory of objects.

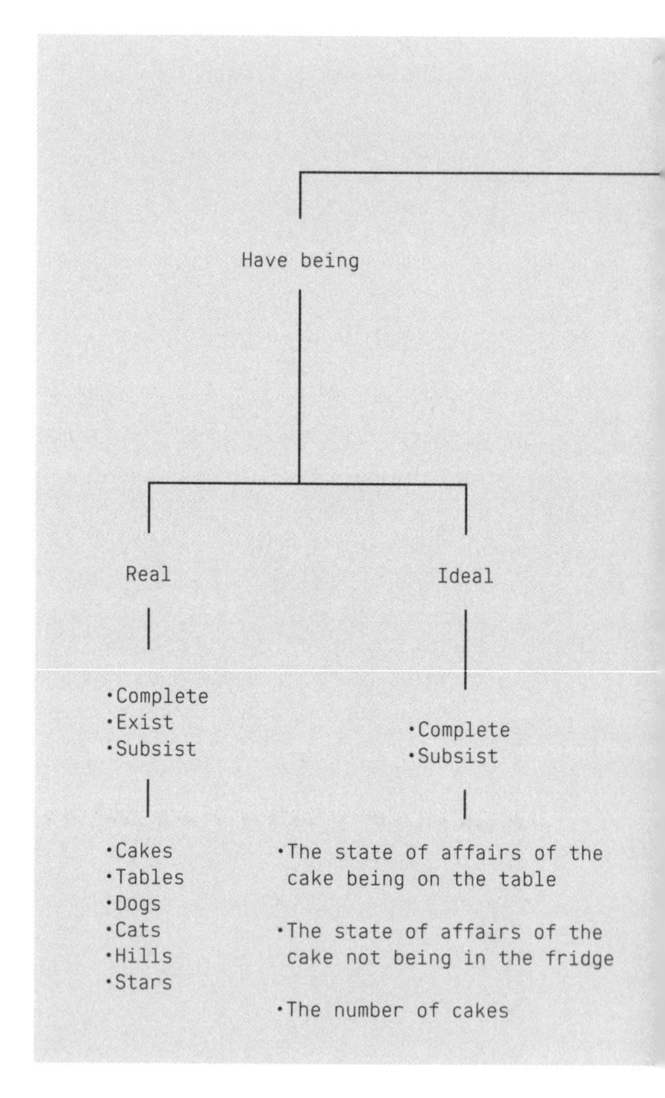

For Meinong, nonactual objects and impossible objects such as perpetual motion machines should be included in any taxonomy of objects. If we can think it, then it exists, somewhere, even if it is in the collective imagination. As Barry C. Smith, director of the Institute of Philosophy at the University of London, remarked in a BBC program about the word "the," "'the' can even have philosophical implications. The Austrian philosopher Alexius Meinong said a denoting phrase like 'the round square' introduced that object; there was now such a thing. According to Meinong, the word itself created non-existent objects, arguing that there are objects that exist and ones that don't—but they are all created by language. 'The' has a kind of magical property in philosophy."[10]

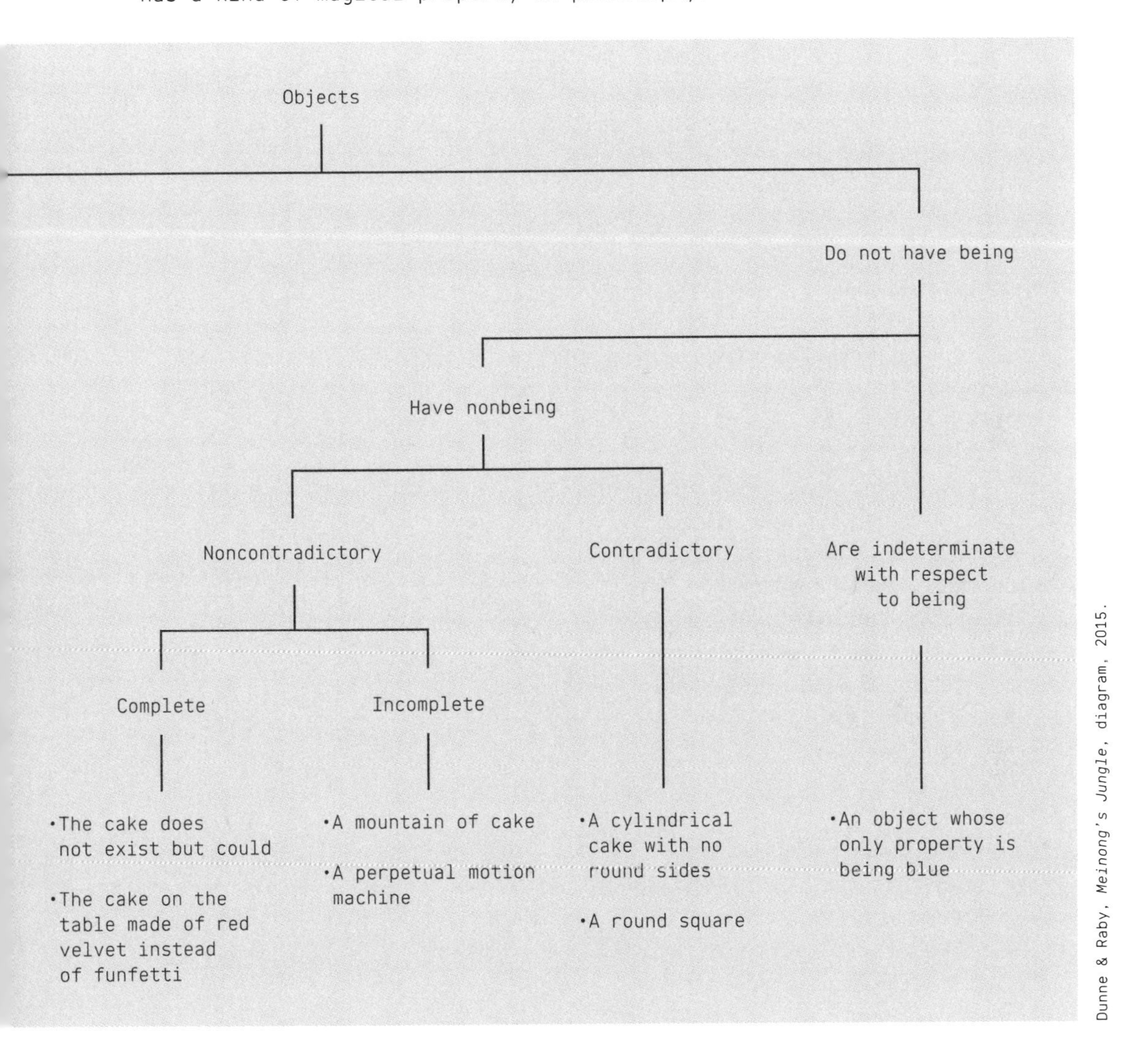

Dunne & Raby, *Meinong's Jungle*, diagram, 2015.

The appeal of his taxonomy is its focus on objects in their broadest sense and their different modes of being in the world: actual entities (e.g., tables and chairs), nonactual entities or abstracta (e.g., numbers and ideas), fictional entities or ficta (e.g., Sherlock Holmes), impossible objects (e.g., round squares), and even unthinkable objects (e.g., having the property of being unthinkable). They may not all be actual, but they do all exist in some way, somewhere, even if it is in the imagination, which is also part of the world. It is this acknowledgment of the many kinds of real that appeals to us. Moving beyond binaries such as real/fictional, and developing a sensitivity to different shades of real when it comes to thinking about the world people live in and the objects designed to aid its inhabitation. Designers say there are real and not real objects, and they want to deal with real ones. Meinong says there are objects that are real, sort of real, hyperreal, not quite real, really real, and so on. We want to design for this world. A world that more fully reflects the range of objects that people interact with daily.

For many philosophers, the theory of objects is a mess; in fact, it is known as "Meinong's Jungle," which in itself is appealing to us. It is a conceptual space where all kinds of strange objects are welcome and even celebrated, and where all are treated equally. And it is the starting point for our interest in "impossible objects." Impossible, because they belong to different reality systems from the present one. At odds with it in ways that produce a conceptually useful wrongness.

Another way of thinking about impossible objects is as "weird" objects in the sense outlined by Mark Fisher in his book *The Weird and the Eerie*: "The weird is a particular kind of perturbation. It involves a sensation of wrongness: a weird entity or object is so strange that it makes us feel that it should not exist, or at least it should not exist here. Yet if the entity or object is here, then the categories which we have up until now used to make sense of the world cannot be valid. The weird thing is not wrong, after all: it is our *conceptions* that must be inadequate."[11] And this is their value.

Fiction as Fiction

Although this book enthusiastically engages with fiction in design and embraces what is usually deemed unreal, it is not about blurring boundaries between the real and the unreal in ways that take us into the realms of fake news, post-truth politics, and other assaults on common sense and rationality. Nor is it an argument for a more ontologically diverse *actual* world where each person lives in their own subjective reality. Rather, it is a call for more ontological plurality *within* design, to expand the design imagination. Using design to materialize proxies for nonactual or imagined worlds through the design of everyday things. We like this quote by author George Saunders: "And that's what fiction does: it causes an incremental change in the state of a mind. That's it. But, you know–it really does it. That change is finite but real. And that's not nothing. It's not everything, but it's not nothing."[12] Fiction does not displace fact in design but instead sits alongside it, giving form to the stories that people tell themselves, and live by and through. Not displacing science, for example, but seeing it as one part of a larger reality that includes stories and other ways of making sense of the world, both actual and imagined.

A Will to Imagine

In *Imagining the University*, Ronald Barnett argues, "Imaginative thinking has to be counter-ideological thinking. This in turn requires a space where such thinking might be fostered. But the space is as much internal to the mind as it is within the structural spaces of the university. The imagination all too easily censures itself, and is complicit in its own diminution. In the end, a will to imagine has to be present. The fundamental tale of the imagination is that it believes in its own right to imagine."[13] We believe this "will to imagine" needs to be more strongly encouraged, facilitated, and celebrated in design. Despite the breadth of topics designers can engage with, which has dramatically expanded over the last two decades or so, there are still many unacknowledged constraints placed on the role of imagination and an alignment with industry norms that are viewed as a benchmark for reality. Imagination in design is too often viewed as a form of escape, unnecessary indulgence, or retreat from reality, even in education. Yet if reality itself has become so problematic, being realistic might not always help; as writer Ursula K. Le Guin asks, "The direction of escape is toward freedom. So what is 'escapism' an accusation of?"[14]

Imagination can also allow designers to momentarily step outside our usual commitments to what we think reality is in order to test out ideas, explore other ways of being, and develop tools for reflecting on the way things are now, and how they might be otherwise.[15] A place of temporary occupancy rather than a permanent abode. As Barnett writes elsewhere in *Imagining the University*, there is always a "tension here between the real and the imagined, between attending to the world as it is and as it might be, between being responsible in and to the world and offering creative and alternative visions of the world."[16] And this for us is the value of the designing for the "not here, not now": objects can simultaneously be in the world as it is through their physicality and apart from it conceptually.

Spiraling Loops of Curiosity

Much has been written about what is wrong with design. And there is certainly much to say on this topic.[17] Arguments have raged since the beginnings of the Industrial Revolution; see, for example, William Morris's fiction, *News from Nowhere* (1890).[18] But there is also much that is good about design—its ability to make abstract ideas, values, and beliefs concrete through the design of everyday objects, for example. This book focuses on design's potential, on how design itself might be otherwise, and in this sense it is optimistic. It advocates for a form of design more concerned with ideas than solutions. Bringing forms of thought usually found in the worlds of literature, philosophy, and the edges of science into conversation with the stuff of everyday life.

At its heart is the idea of the "design proposal," or proposition. But as a prompt rather than prescription. Like the most compelling utopias, they are not destinations to be aimed for but rather, are intentionally unrealizable—impossible by design, serving instead to nourish the creative, intellectual, and imaginative ground from which new possibilities, still unknown, might begin to emerge.

The essays, experiments, and projects in this book evolved in dialogue with each other through what we like to think of as a designerly form of inquiry: spiraling loops of curiosity consisting of questions, experiments, classes, readings, conversations, lectures, projects, writings, and reflections. They draw on an eclectic range of ideas from fields including philosophy, anthropology, literature, physics, and fine art. We do not claim to treat or handle these ideas as they would be in their home disciplines; our focus is purely on exploring their usefulness for thinking differently about the design of everyday objects and our interactions with them.

Each section looks at the idea of the "impossible object" through a different lens or different perspective—varieties, scales, uses, and misuses, places of encounter, histories, fiction, and language. We also explore ways of writing and thinking not only about objects but also through objects—actual and imagined, found and designed, made from words, images, and matter. This allows us, the authors, to design (and write) from the position of what Le Guin has called "realists of a larger reality."[19] A reality that fully embraces the imagination and all that is yet to exist, or might never exist—what we currently think of as unreality, or the "not here, not now."

2 An Arch
Impossibl

ive of
e Objects

Sewing machine, A. F. Johnson, January 13, 1857. Boston, MA, patent no. 16,387. Division of Home and Community Life, National Museum of American History, Smithsonian Institution.

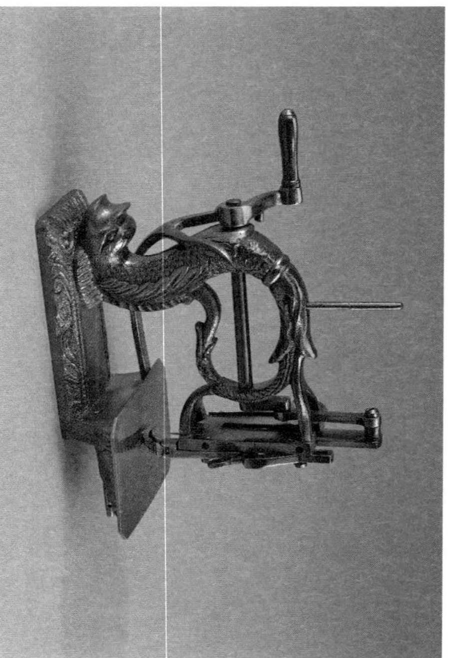

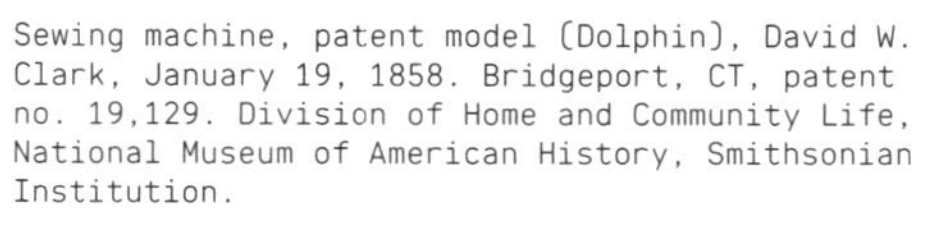

Sewing machine, patent model (Dolphin), David W. Clark, January 19, 1858. Bridgeport, CT, patent no. 19,129. Division of Home and Community Life, National Museum of American History, Smithsonian Institution.

Sewing machine, George Hensel, July 12, 1859. New York, NY, patent no. 24,737. Division of Home and Community Life, National Museum of American History, Smithsonian Institution.

Chapter 2

An Archive of Impossible Objects

For much of the 1800s, the US Patent Office in Washington, DC, required a model as well as a written description and drawings to be submitted for every invention registered. The idea was to enable someone with some technical skills and knowledge to make use of the device. Inventors, patent agents, and even the public could visit and request an object for study. The models were also a way to "educate the public, stimulate creativity and enterprise, and inspire a sense of America's destiny." Over time, this repository of material proxies for possible worlds expanded to become a kind of public library of things. Numbering less than 25,000 models in 1856, it grew to 100,000 in the late 1860s, and continued to grow until the practice ended in 1870 with approximately 250,000 models.[1]

Old Patent Office, model room, 1861–65. Prints and Photographs Division, Library of Congress, [LC-DIG-cwpbh-03283].

Could a place like this exist today? A place for collecting, juxtaposing, studying, and encountering objects belonging to different (impossible) worlds, or "reals," found and designed, actual and imagined.[2]

On the surface, this might sound a little similar to a cabinet of curiosities, or *wunderkammer*, but it has a few important differences. Wunderkammers present their strange contents with a view to shock or surprise, although their eclectic and nonhierarchical organization is appealing. There are also a number of well-known museums like the Museum of Jurassic Technology, a favorite of ours, that serve as architecture-scale cabinets of curiosity. But part of the Museum of Jurassic Technology's appeal or charm is its conscious blurring of boundaries between the real and the unreal, attempting to sneak the unreal into the real. Ideally, the exhibits in an Archive of Impossible Objects would clearly belong to worlds that are unreal by the standards of the current real. And exist simply to aid contemplation and reflection through contact with other systems of thought, or reals, made concrete in the form of designed objects. It is probably closer in spirit to Jorge Luis Borges's "Chinese encyclopedia" quoted by Michel Foucault in the introduction to *The Order of Things*.[3]

(Where) Animals are divided into:

(a) belonging to the Emperor
(b) embalmed
(c) tame
(d) suckling pigs
(e) sirens
(f) fabulous
(g) stray dogs
(h) included in the present classification
(i) frenzied
(j) innumerable
(k) drawn with a very fine camelhair brush
(l) *et cetera*
(m) having just broken the water pitcher
(n) that from a long way off look like flies

Foucault suggests that "in the wonderment of this taxonomy, the thing we apprehend in one great leap, the thing that, by means of the fable, is demonstrated as the exotic charm of another system of thought, is the limitation of our own, the stark impossibility of thinking *that*."[4]

Like this taxonomy, impossible objects are seemingly incompatible with existing sensemaking frameworks, or at least they make a different kind of sense. But they are not nonsense objects or Dada-like gestures challenging the very notion of sense. They are invitations to question the frameworks we use to make sense of the world and even what a "world" is. A place for exploring different kinds of thought that stretch the imagination in ways that not only help imagine how things might be otherwise but also how the way we think about the metaphysical work in progress we call reality might change too. Impossible in this context means within our own narrow sense of what we deem real or unreal. What can and cannot exist within our specific ontology.

A collection of "impossible objects" might also serve as a vehicle for reflecting on how limits to what is possible are constructed, both within us individually, and in our collective imaginations and the role design plays in this. Where do these boundaries come from, and how are they internalized?

Poet and musician Sun Ra once remarked in an interview, "The impossible attracts me...because everything possible has been done and the world didn't 'change."[5] This is a painfully counterintuitive idea for design, where possibility is nearly always the goal. If something is impossible, then it is unrealistic and therefore pointless, a waste of time, valueless. But from our perspective, being possible simply ensures that something fits existing reality and thus reinforces it. A reality that for many people no longer works. Impossible objects resist being assimilated through the sensemaking schema of a problematic reality. Their strangeness highlights the schema used to decide what can and cannot exist in this world, even in the imagination, exposing them to critique and opening the way for other possibilities. Impossible objects attempt to put the complex relationship between the possible and the impossible into a state of imaginative superposition, neither one nor the other, nor both. Hovering in a not here, not now.

History also offers many helpful perspectives on this, specifically intellectual shifts that allowed new objects (of knowledge) to pass in and out of existence.[6] This might be due to breakthroughs in understanding or the abandonment of outmoded worldviews, such as those explored by Darren Oldridge in *Strange Histories*. On beliefs about the afterlife, for example, he writes:

> The close scrutiny of the afterlife by pre-modern scholars strikes many people today as shockingly "literal minded." This is a telling

> phrase. For the great majority of people in pre-industrial Europe, as well as the colonies of seventeenth-century America, there was nothing allegorical about life after death. The heaven they believed in was real. In the same way, modern scientists assume that the objects they examine—atomic particles, chemicals, gasses or planets—really exist. They are literal indeed too. What is truly remarkable about medieval thinkers like Aquinas is their complete acceptance of the reality of heaven and hell, angels and demons. Once this belief is granted, the questions they asked can be accepted as rational and important. If human souls survive death, it is reasonable to ask where they go.[7]

The archive might serve other purposes too, encountering ontologies for pure pleasure, for instance. As Thomas G. Pavel suggests in *Fiction and the Ontological Landscape*, these might include delightfully intriguing categories such as "discarded ontologies," "ontological ruins," and "ontological relics." To this we can add nonhuman ontologies that by their very nature are impossible to grasp for human-shaped minds. The archive could also serve as a sort of "ontological training ground... to train the members of the community in such abilities as rapid induction, construction of hypotheses, positing of possible worlds, etc."[8] In the context of design, it could serve as a resource for moving beyond futures as the primary way of framing the "not here, not now." Essentially, a place that celebrates the ontological imagination.

It always strikes us how odd it is that, as design educator Matt Ward put it in "Design, Fiction and the Logic of the Impossible," "We are a discipline that is reliant on our creativity and imagination, but have become terrified of the imaginary."[9] Other disciplinary imaginations go far beyond what design deems permissible, showing that reality may not be as realistic as we think. There are quantum physicists who entertain the idea that new worlds branch off from this one each time a measurement is made, astrophysicists who believe that this world is but one among an infinite number of universes, and philosophers who engage in possible worlds talk where, in one world, there are indeed unicorns.[10] We believe design has much to learn from these fields, and that environments are needed to explore this more, putting design into conversation with philosophy, speculative forms of science, and literature. Fields that have a long history of working with ideas that stretch the imagination and allow us, as designers, to see beyond the constraints of existing realities.

In design, when people hear the term "impossible object," they might think of Jacques Carelman's *Catalogue d'Objet Introuvables* (1969) and similar objects that play with functional logic. Exploring through many examples and innumerable permutations, highly poetic ways of negating a familiar object's functionality. For example, a "tandem convergent," where the cyclists face each other. These mirror philosophical definitions of "impossible" based on situations where the laws of logic fail. The viewer understands the logic of a bicycle and a design that denies its core function. They are physically embodied logical contradictions.

A more interesting example is the "rabbit-duck illusion" made famous by Ludwig Wittgenstein in his *Philosophical Investigations* (1958). Although usually treated as an image, the physical version is even more enjoyable to behold; see, for instance, artist Simon Cunningham's *Duckrabbit* (2019).

Simon Cunningham, *Duckrabbit*, 2019.

It is not just the animal, a duck-rabbit, that is impossible but also the ability for the human brain to see the duck-rabbit—that is, a duck and rabbit simultaneously. You can only see either a duck or rabbit at any one time. Then there are examples like the *Ambiguous Cylinder Illusion* by Kokichi Sugihara.[11] Although illusions like this really only work when viewed as a video of specific reflections of an object from a precise viewing angle. A sophisticated form of anamorphism. But these examples are a little too technical and straightforward.

Just Because Something is Impossible, Does Not Mean It Is Not Possible

There are thinkers who argue that the impossible cannot even be conceived. Or that once it is thought or imagined, it ceases to be impossible, or like René Descartes's one-thousand-faced geometric figure, it can be intellectually grasped but not imagined.[12] Philosopher Tristan Garcia believes that "possibility includes impossibility, since impossibility is possible in the sense that it is something (even if it is a word, a contradiction, a mistaken idea and nothing more). Impossibility is not opposed as a whole to possibility, but marks one possibility among others."[13] Others claim that logical contradictions such as the "principle of the excluded middle," which allows statements to be only either true or false, or the "principle of noncontradiction," which states that nothing can be true and false at the same time, are actually unthinkable.

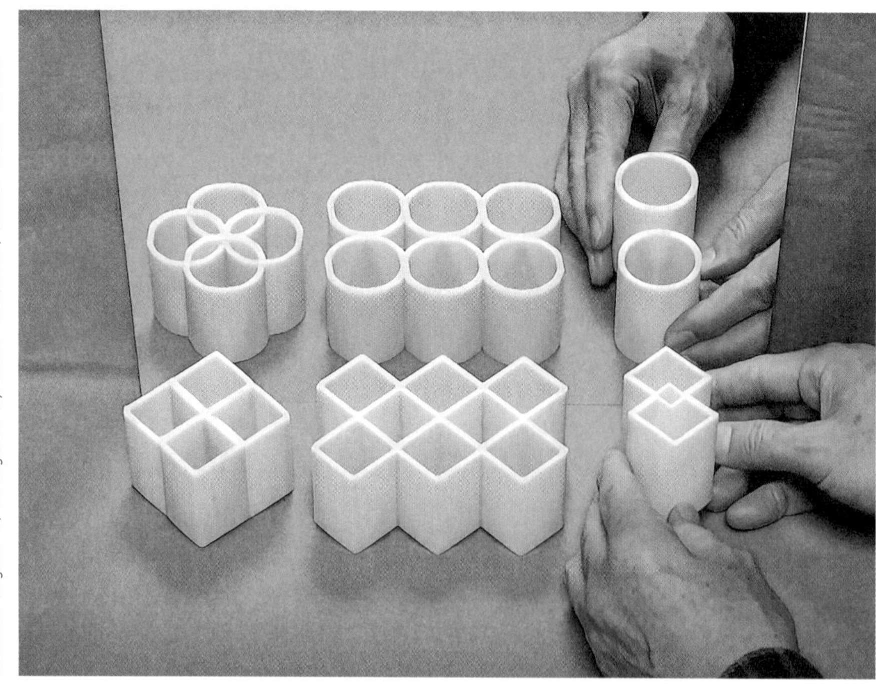

Kokichi Sugihara, *Ambiguous Cylinder Illusion*, 2016. Video still.

But as philosopher Graham Priest notes in his essay *Beyond True and False*, impossibilities like these are limited to Western thought and logic. He points out that in fifth century BCE in India, for example, rather than two opposing states for any statement—true or false—there are four, based on the "catuṣkoṭi principle," which means four corners: "There are four possibilities regarding any statement: it might be true (and true only), false (and false only), both true and false, or neither true nor false." Priest goes on to explain how thought like this can be possible in Western systems of logic too, even if untypical. For instance, Aristotle claimed that "contingent statements about the future, such as 'the first pope in the 22nd century will be African,' are neither true nor false. The future is, as yet, indeterminate."[14] And an example of something being both true and false is the well-known liar paradox: "This statement is false." If it is true, then the statement is false, and if false, then it is true, seemingly being both true and false.

Priest also describes a fifth state, to which none of the four states of catuṣkoṭi apply: ineffability. He suggests this concept was used to prevent certain ideas from being drawn into conventional reality. If they were deemed ineffable, then they could be conceptually located just beyond the boundaries of human thought and imagination, generating a quandary for Western philosophy. How, for example, can we know something unknowable without saying it and therefore making it knowable? Is this yet another kind of impossible object—the "Buddhist paradox of ineffability"? This truly impossible object is not one we can ever encounter, at least in this archive, and is best left to philosophers, although it does usefully mark one end of a spectrum.

At the other end, in contrast to these philosophical reflections on impossibility, lies other kinds of impossibility entangled with materiality, and that is the focus of this archive. Although the material world might not be quite as solid as we like to think either. In his short text of 1966, *On Fiction*, Vilém Flusser explores how something as seemingly obvious or even banal as a piece of furniture can appear to be a contradiction, an impossible object: "Take, as an example, this table. It is a solid board that my books lie upon. But, as we know, this is a fiction. This fiction is referred to as 'the reality of the senses.' If considered from a different perspective, the table is a practically empty electromagnetic and gravitational field over which other fields called 'books' float. But, as we know, this is a fiction. This fiction is called 'the reality of exact science.'"[15]

But Why Objects?

At a time when so much attention is being directed at services, systems, augmented reality, the metaverse, virtual reality, machine learning, and artificial intelligence, it might seem a little strange to focus on physical things. But for us, objects, things, the stuff of everyday life, are still a powerful medium for engaging the imagination. They can serve as conceptual anchors tying ideas from elsewhere or belonging to different reality systems to our own material reality. Foucault's brief description of a heterotopia in "Of Other Spaces, Heterotopias," a lecture he gave in 1967, captures this quality nicely:

> I believe that between utopias and these quite other sites, these heterotopias, there might be a sort of mixed, joint experience, which would be the mirror. The mirror is, after all, a utopia, since it is a placeless place. In the mirror, I see myself there where I am not, in an unreal, virtual space that opens up behind the surface; I am over there, there where I am not, a sort of shadow that gives my own visibility to myself, that enables me to see myself there where I am absent: such is the utopia of the mirror. But it is also a heterotopia in so far as the mirror does exist in reality, where it exerts a sort of counteraction on the position that I occupy.[16]

When we first began to write about designed objects as a critical medium in the 1990s, we struggled to find frameworks for thinking about objects beyond semiotics and material culture, but today, despite or maybe because of a move toward dematerialization, there seems to have been a resurgence in theories about things and objects. Just a few:

Theories about objects so large we can only ever grasp one small part of them (hyperobjects).[17] Theories on how objects of scientific study move in and out of existence over time, and how once matter and meaning fuse, they speak (object biographies).[18] Theories about how objects are always entangled with complex networks of actants, both human and nonhuman (actor network theory).[19] Theories about how humans and objects interact in literature and culture (thing theory).[20] Theories about how intra-actions, rather than simple interactions, bring objects into existence (agential realism).[21] Theories about object autonomy and their independence from the human mind (object-oriented ontology).[22] Theories that recognize that both human- and nonhuman-related objects have agency and shape reality (vibrant matter).[23] Theories of objects as placeholders or mental representations for an unknowable reality much like the icons on our computer desktops (conscious realism).[24] Theories that look beyond language to materiality as a primary source of meaning (new materialism).[25] Theories forced to conclude that objects can simultaneously exist in several states and even universes (quantum mechanics).[26] Theories that claim all matter has consciousness of sorts (panpsychism).[27] And theories where objects belong to many different worlds, each with its own ontology (ontological turn).[28]

In this context, an object is a slippery thing indeed. Its conceptual leakiness and material indeterminacy make for a rich and complex cloud of possibilities when fused with the idea of other reals. There is something special about encountering a physical proxy for another reality, or world, in an unmediated one-to-one setting and engaging with it as a thing in space. These objects are not behind screens or the glossy surface of a photo but instead in the room with us, coexisting, sharing space, appealing to all of our senses, not just our eyes. Objects or things materialize only small parts of imagined worlds (or different reals), unlike architecture or science-fictional cinema, where whole worlds can be represented. Ideally, this fragmentary approach leaves more room for the viewer to imagine a world and its system of reality for themselves. Objects are suggestive and open-ended. Prompts for the imagination rather than fully formed realities. An approach we have called "world hinting" in our book *Speculative Everything* (2013).

But let's visit the archive.

Dunne & Raby, *An Archive of Impossible Objects*, 2023. Computer-generated image: Carolyn Kirschner.

Object 1: An Object Made from Words

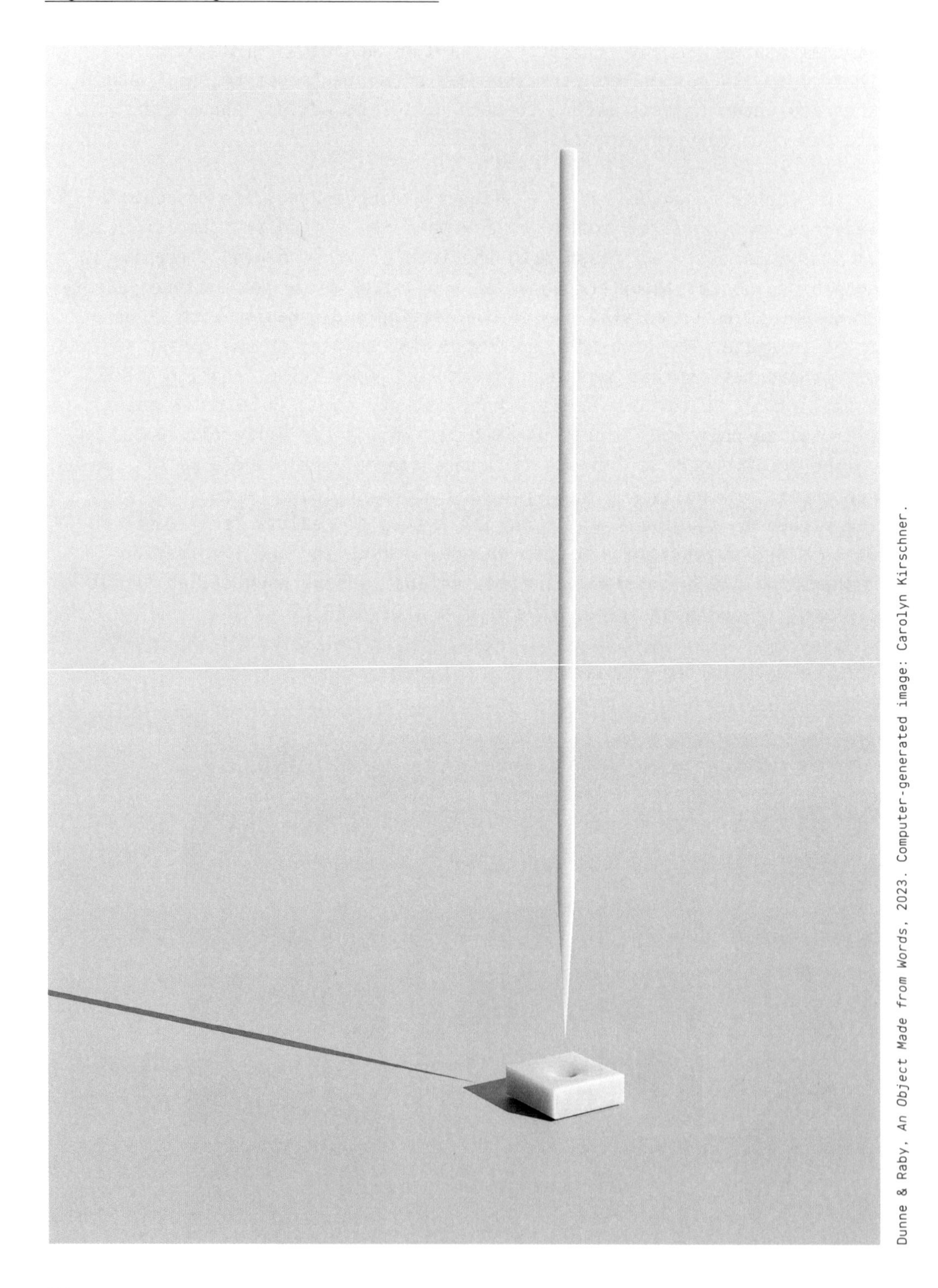

Dunne & Raby, *An Object Made from Words*, 2023. Computer-generated image: Carolyn Kirschner.

In his short story *Sylvan's Box*, Graham Priest narrates a visit to the recently deceased Richard Sylvan's house.[29] While there, he discovers a box labeled "Impossible Object," and on opening it, is surprised to see that it is both empty and contains a statue. After some speculation with his friend Nicolas Griffin on its possible meanings, rather than sharing this discovery with the world, they decide to dispose of it. Priest drives off with it in his car, while Griffin buries it in the garden.[30] The story is an example of how certain kinds of contradictory objects that cannot exist as actual, physical objects can be thought or imagined into existence through words. To attempt to materialize this example as an actual object would probably produce little more than a banality. Another more extreme instance from writer Haruki Murakami's *First Person Singular* can never be realized: "'Think about it,' the old man said. 'Close your eyes again, and think it all through. A circle that has many centers but no circumference.'"[31]

Objects like these interest us because being made from words, like all materials used to make things, they lend their objects specific qualities, whether poetic, conceptual, aesthetic, or practical, that are not always achievable through other materials. Being avid readers of fiction, especially science fiction, we are always on the lookout for new and interesting objects that celebrate their wordy materiality. There are surprisingly few. Or at least they are hard to find. Of course, science fiction is full of novel and inventive textual objects, but they tend to go with the grain of reality, reinforcing the conceptual foundations of whatever world or system of reality they are part of. But occasionally one does encounter objects made from words that like *Sylvan's Box*, challenge the specific reality they appear in.

In Arkady and Boris Strugatsky's novel *Roadside Picnic* (1972), overnight, six zones appear scattered across earth containing objects that don't quite align with the laws of physics. One, for example, is a "Full Empty" consisting of two disks held about 450 millimeters apart by an unknown force with an incredibly heavy liquid of unknown purpose suspended between the disks. A simple device, but one that again, if it were materialized, would be far less interesting than its textual version.[32] Another, more poetic example can be found in Flan O'Brien's *The Third Policeman*, where in one scene, a spear is introduced with a point so sharp that it penetrates the body by several inches before the visible part makes contact and draws blood. The suggestion is that the point has been sharpened down to an atom so it can pass through the space between the atoms in a person's hand, "He kept putting it near me and nearer and when he had the bright point of it about half a foot away, I felt a prick and gave a short cry. There was a little bead of my red blood in the middle of my palm."[33]

Objects made from words are also regularly put to work in thought experiments, where they are almost surgically isolated from their context to clearly illustrate or test an idea. The most interesting involve objects that test different kinds of paradoxes using specially "designed" impossible objects made from words. For instance, *The Ship of Theseus*: If over time every component on a ship was replaced, to the extent that all original elements were no longer present, would it still be the same ship? If not, at what point did it cease to be the same object? Here, an object made from words is used to uniquely explore ideas around identity in relation to things over time. Some have argued that this puzzle tells us more about the human mind and common intuitions than about actual things in the world. A related and more contemporary example is Derek Parfit's teletransporter. A person steps into a teleport on earth, is analyzed and measured down to the atomic level, this information is then sent to Mars, and the person is reassembled. Considering the person would remember stepping into the machine and everything else about the experience, did the original person travel to Mars? Again, it is something best explored in text; the machine is another impossible object made from words.[34]

These kinds of objects work best when they are used to test an idea, especially one that is seemingly paradoxical and sheds light on language's complicated, sometimes troubled relationship to reality, rather than being a textual substitute for something that could exist in reality or be visually represented, which would just be a description or fictional object as opposed to being an impossible object made from words.[35]

But what does this mean for design? It suggests to us that there is a place for things made from words, not as a shorthand for what could actually be designed, or as stories that focus on designed objects and their role in our lives, but as objects in themselves that function best as texts doing things that other objects made from matter, existing in space-time, could never achieve.

But we are designers of physical things.

Object 2: A Machine-Generated Impossible Object

Dunne & Raby, with Philipp Schmidt, *A Machine-Generated Impossible Object*, 2018.

If you do an online search for "impossible object," the screen will quickly fill with images of a specific kind of object defined in Wikipedia rather technically as "a type of optical illusion consisting of a two-dimensional figure which is instantly and subconsciously interpreted by the visual system as representing a projection of a three-dimensional object although it is not geometrically possible for such an object to exist (at least not in the form interpreted by the visual system)."[36] A visual version of the kind of linguistic contradictions mentioned in the introduction to this section. Examples include Escher-like two-dimensional representations of three-dimensional spatially contradictory scenes and other familiar objects such as Necker cubes and variations on the impossible fork (also known as an impossible trident, devil's tuning fork, or blivet).

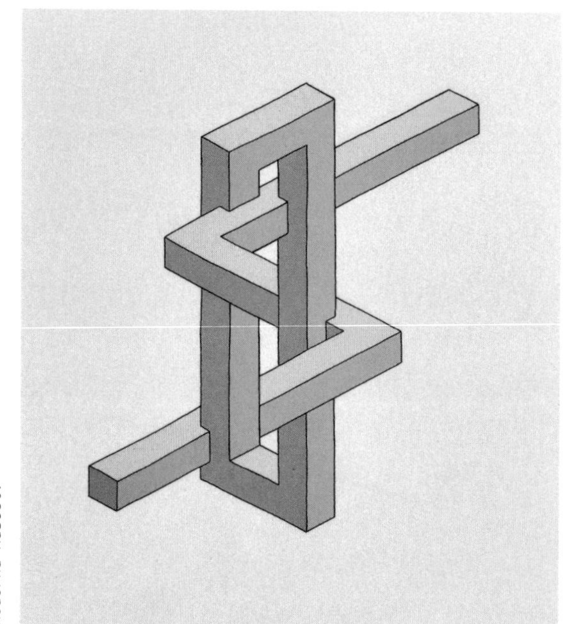

Oscar Reutersvärd, 326 *gkgk*. Donation, 1983, from Gunnar Berefelt, Moderna Museet.

These form a foundational spatiovisual grammar for constructing more elaborate scenes such as buildings, furniture, and in the case of the highly successful computer game *Monument Valley*, a whole world. Although M.C. Escher and Roger Penrose are most strongly associated with this field, it is Swedish graphic artist Oscar Reutersvärd who pioneered the area and developed an extensive body of work over his career, creating approximately four thousand examples of what he called "Impossible Figures."[37] Many of these figures use a similar visual grammar, and we wondered if we could use machine learning to add a new example to the catalog of spatial impossibilia or maybe even to Meinong's taxonomy of objects.

Working with research assistant Philipp Schmitt in summer 2018, we fed a generative adversarial network (GAN) a series of images based on familiar illusions–Escher-like staircase details, variations on the impossible trident, and so on.[38] Over time, what looked to us like hesitant sketches of vaguely impossible objects slowly began to emerge before collapsing into a dense series of impenetrable marks. We collated the GAN's efforts in a sketchbook we called *In Search of an Impossible Object*. Each page shows a selection of attempts at visualizing a new impossible object. None succeeded. But collectively, they suggest something lying just beyond what can be represented graphically. It is as if a machine, for a moment, attempted to show humans, in a visual language we would understand, a side to themselves that is permanently unknowable to us.

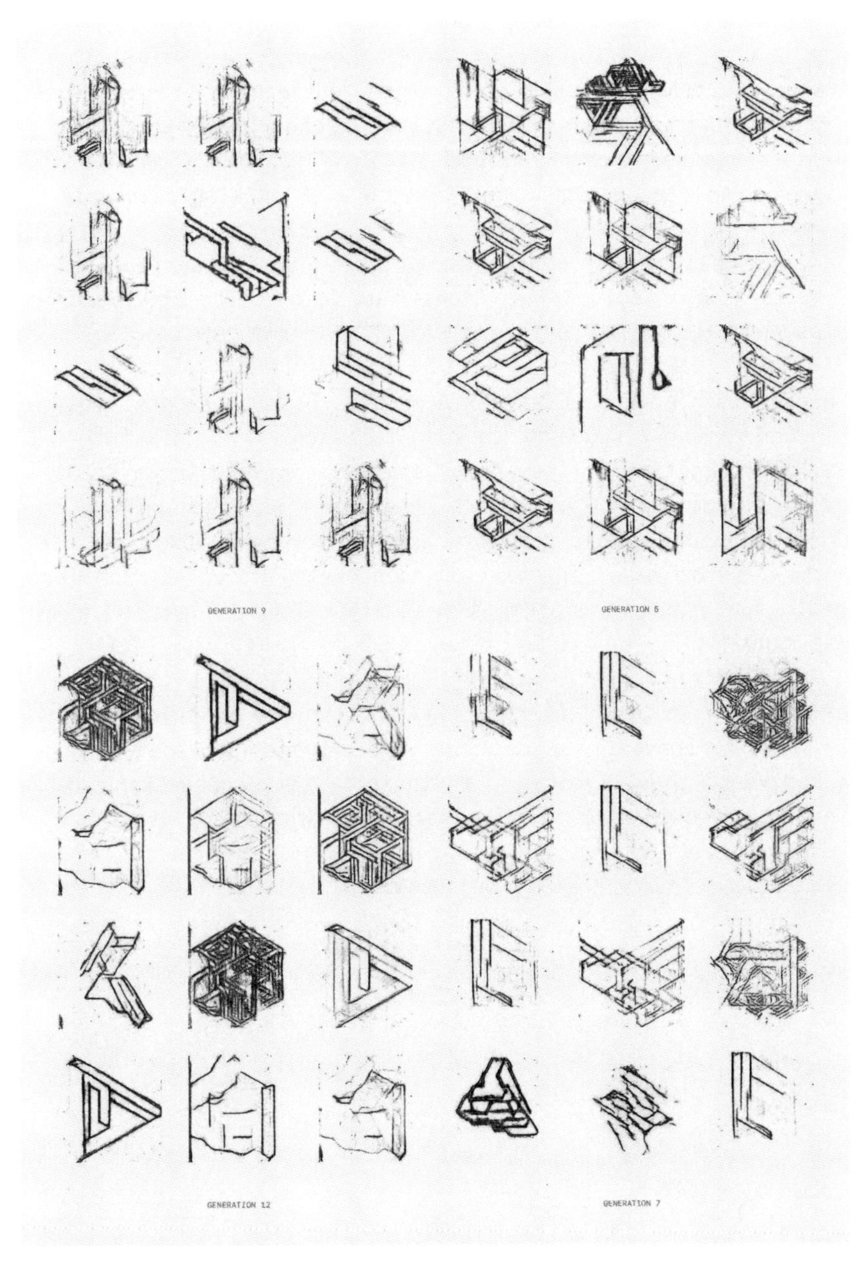

Dunne & Raby, with Philipp Schmidt, *In Search of an Impossible Object*, 2018. Page spreads.

Looking at the generated images, all derived from low-resolution original material, it is striking how similar they appear to marks made by a human intuitively feeling their way toward an idea. Some look as though they are preformed ideas extracted from its machine imagination, or more accurately, the multidimensional space within its neural network where all possible impossible objects might already exist, unrealized. Others look as if the idea is somehow already on the page, being teased out through the marks.

This is of course a simple and rudimentary experiment done before software like DALL-E and Midjourney became publicly available, but whereas many machine learning experiments focus on imagery, paintings, and other purely visual material, resulting in sometimes quite striking painterly outcomes, we wanted to see what would happen when it was applied to objects, or at least the visual grammar and syntax of their two-dimensional representation, in lines. Could a GAN suggest new forms of object logic or spatial arrangements lying just beyond what we as humans can imagine?

We began to wonder if a more sophisticated process using more powerful machines would generate new kinds of optical illusions that hack into our perceptual and cognitive processes for spatial analysis in ways not previously imagined–impossible objects with a specific function. As with all optical illusions, could they tell us something new about how the human mind works–in this case, in relation to geometry? Could a machine become a partner for generating prompts that might lead to new insights into the human mind?[39]

To do this effectively, though, it might be necessary to go beyond the use of models based on conventional kinds of intelligence and instead work with more neurodiverse intelligences–a reminder of yet another kind of bias built into these programs and that there are many other, yet to be explored forms of intelligence.

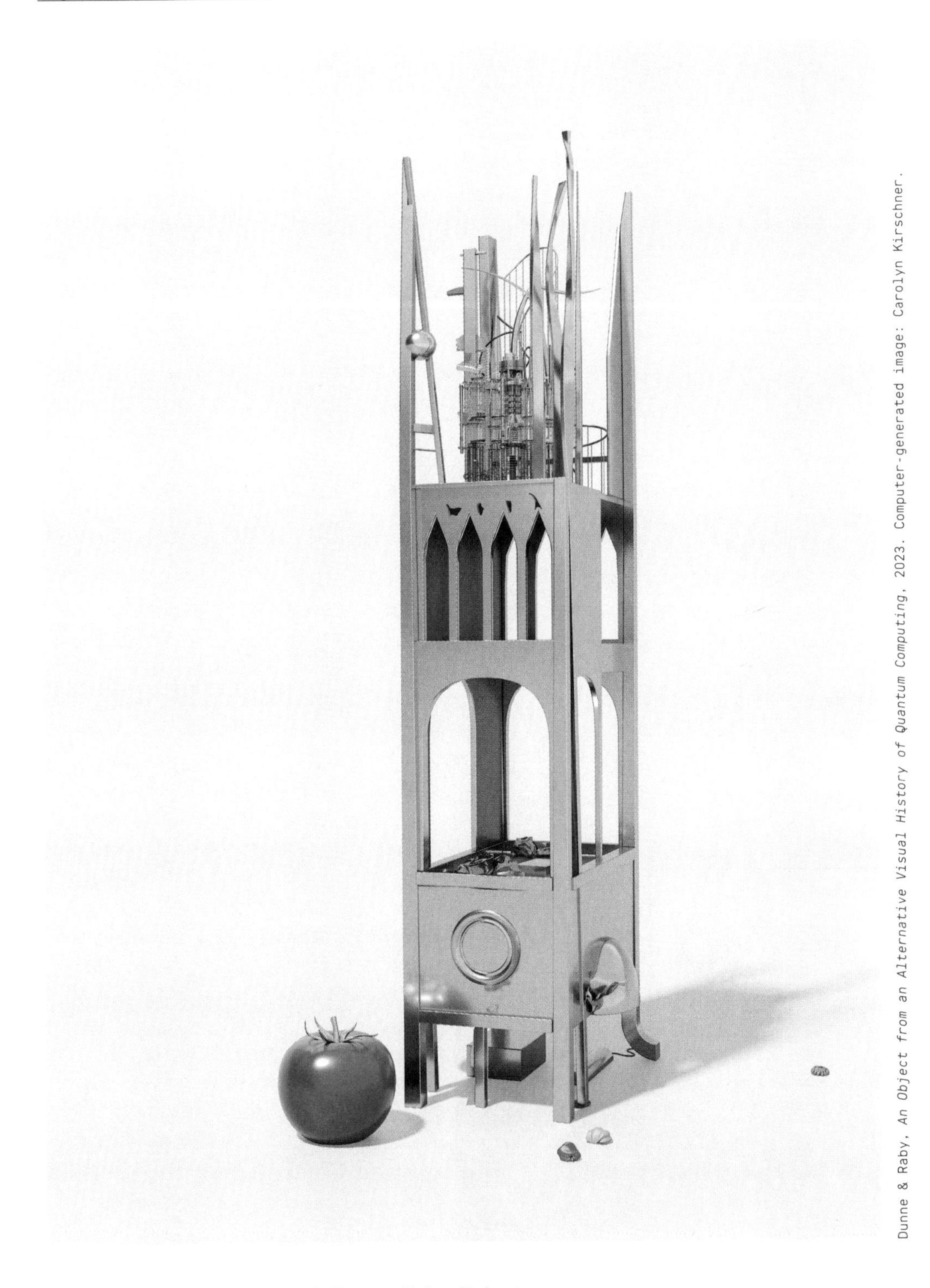

Dunne & Raby, *An Object from an Alternative Visual History of Quantum Computing*, 2023. Computer-generated image: Carolyn Kirschner.

Quantum mechanics is of course home to many impossible objects, including one of the most famous, Schrödinger's cat, which is both alive and dead.[40]

But where to even start.

For us, the quantum computer is a fascinating point of contact between the worlds of quantum mechanics and everyday materiality. In a quantum computer, somewhere, seemingly magical objects can exist. Objects or particles that can be in two places at once, or exist in two states at once—so-called cat states. These entities are qubits, and unlike classic bits that exist as either zeros or ones, qubits are both zero and one and all states in between. A quantum computer is a zone where quantumness becomes physical in this reality. Well, sort of. A quantum computer is a real thing, just about. We say "just about" because in order to work, it needs to be practically isolated from physical reality to prevent it from interacting with the environment. Any interference at all can cause a qubit to collapse from a state of multiple possibilities into one singular state, a process known as decoherence, causing it to lose any information it has stored.

These devices make use of ideas that contradict many of our most common beliefs when it comes to making sense of reality. It exists despite reality, not because of it.

Quantum computing, it is hoped, will eventually help solve problems currently considered impossible in fields like cryptography or modeling complex molecular interactions to develop new drugs. But for some of the early conceptualizers of quantum computing, it was more like a philosophical machine. In the 1980s, physicist Richard Feynman thought that to truly understand and model natural processes, which are essentially quantum, some kind of quantum computer would be necessary. While David Deutsch, an advocate of the many-worlds interpretation (MWI) of quantum mechanics and pioneer of some of the ideas behind quantum computing, believed that a quantum computer would eventually reveal the existence

Dunne & Raby, images generated with the assistance of DALL-E 2 using the prompt "paintings in the style of... of people using a quantum computer," 2023.

of many worlds and prove it true.[41] Showing that Schrödinger's cat, for example, is indeed both alive and dead, but in different worlds. Quantum computers are fast because they entertain all possible outcomes simultaneously, and if like Deutsch you believe in the MWI of quantum mechanics, then possibly, they are calculating in many different parallel realities at once. We will return to this in the next chapter, "Quantum Common Sense: New Images, Concepts, and Metaphors?"

Struggling to visualize something quantum, we turned to DALL-E 2. What would it generate from the prompt "people using a quantum computer." We were disappointed to discover that it didn't seem to know what a quantum computer looked like. By adding more prompts based on different historical painting styles and artists, we made more progress, and it began to generate images that suggested a strange alternative visual history of quantum computing reflected in various historical styles of painting.

Surprisingly, or maybe not, the most quantum computerlike object appeared in a machine-generated image in the style of Hieronymus Bosch's painting *The Garden of Earthly Delights*.

Dunne & Raby, image generated with the assistance of DALL-E 2 using the prompt "a painting of a quantum computer in *The Garden of Earthly Delights* by Hieronymus Bosch," 2023.

Object 4: Swatches of Forbidden, Chimerical, and Imaginary Colors

Dunne & Raby, Swatches of *Forbidden, Chimerical, and Imaginary Colors*, 2023. Computer-generated image: Carolyn Kirschner.

Self-luminous red, reddish green, stygian blue, and hyperbolic orange are all examples of impossible colors, a category that also includes chimerical, forbidden, imaginary, nonrealizable, and nonphysical colors. Color, an essential part of almost every designed object, has its own kinds of impossibility.

There are two main categories of impossible color: those that lie just beyond human perception and are impossible to see, but exist, and those that do not exist physically, but we can "see" in the mind. Although whether any color physically exists is arguable. But what we mean is that wavelengths of a specific length can hit the eye's cones and rods from outside the body and produce a known color in the mind.

During a brief visit to CERN in 2022, an astrophysicist we were talking to mentioned a "microwave-colored sky" when describing the background radiation she spends her time studying. Like most of the electromagnetic spectrum, including radio waves infrared through ultraviolet and X-rays, microwaves have a wavelength that our eyes cannot detect. Visible light makes up a relatively small part of the spectrum located between the infrared and ultraviolet segments, or 400- to 700-nanometers wavelength. Other creatures can see frequencies lying beyond this, maybe even enjoying microwave-colored skies.

There are also what can be thought of as "superlative colors." Blackest black, whitest white, and pinkest pink are just a few. They both physically exist and can be seen. Impossible until recently, but now possible due to technological developments. Vantablack, for example, the blackest black ever produced, absorbs 99.96 percent of light and lends any object whose surface it covers the appearance of a mini black hole. Especially when photographed, where it appears as a two-dimensional image sitting on the surface of the photograph, a semi-impossible object of sorts. Or ultrawhite, the whitest paint ever made, reflects 98 percent of sunlight and may lead to new ways of saving energy. Its inventor, Professor Xiulin Ruan, claims that "every 1% of additional reflectance equals 10 watts per meter squared less heat from the sun...[We] estimate we would only need to paint 1% of the Earth's surface with this paint–perhaps an area where no people live that is covered in rocks–and that could help fight the climate change trend."[42]

Another category of impossible colors that does not physically exist but can be "seen" are chimerical or forbidden colors, hovering between perception and conception in a state of chromatic superposition. Self-luminous red, for example, appears when staring at a white sheet after looking at a patch of green. It appears to glow even though it is notactually there, so cannot emit light. Stygian colors are dark and saturated, but not actual. For instance, stygian blue appears while staring at a black surface after looking at yellow for a while. There are also colors, such as hyperbolic orange, which appears when one stares at an orange surface after staring at a self-luminous orange surface produced on a white sheet after looking at cyan. An impossibly saturated orange.

There are fictional colors too, which cannot be seen and do not exist. They are simply colors that appear in fiction, usually novels. And like the impossible objects in the first section of the archive, they are made purely from words: subred, actinic, ultraindigo, amarklor, garrow, infrawhite, ocarina and so on.[43] A favorite of ours is "Hooloovoo" from Douglas Adams's The Hitchhiker's Guide to the Galaxy, "a super intelligent shade of the color blue," mentioned in David Toomey's wonderful book Weird Life: The Search for *Life That Is Very, Very Different from Our Own.*[44]

Then there are the optical illusions that occasionally circulate on social media as memes, such as "the dress," a kind of a chromatic duck-rabbit that became an internet meme in 2015. Some people saw it as an image of a black and blue dress (which the original dress was), while for others it appeared to be gold and white. Images like this highlight how people see color differently, providing new material and insights for neuroscientists exploring color vision and how perception works.

Each of these colors has applications of one kind or another–creating intrigue as part of a plot in fictional stories, helping researchers understand how the brain works, or reflecting the sun's heat as part of an imagined technosolutionist fantasy. But one of the first impossible objects we encountered, which was presented as such, was an "object with only the quality of being blue" in a lecture about Meinong's theory of objects. At the time, we were thinking about world-building along with the tendency to simply cut and paste preexisting political ideologies and economic systems from one culture or historical period to another–communism, feudalism, authoritarianism, neoliberalism, anarchy, monarchy, planned or market economies, and so on. An object with only the quality of being blue was an invitation to imagine a world where an object like this could exist, not only its physical world, but the story of that world–its cosmology.

At first it might appear to be another fictional color made from words, but we know what blue is. This blue is a special kind of blue made from ideas. By drawing more on philosophy than science, technology, history, economics, or politics, an "object with only the quality of being blue" suggests a very different approach to world-building. There are not so many examples of this, unsurprisingly, but one that stands out is Borges's "Tlön, Uqbar, Orbis Tertius" (1945), a story set in a world informed by Irish philosopher George Berkeley's subjective idealism, where nothing exists independently of the mind:

> The literature of this hemisphere (like Meinong's subsistent world) abounds in ideal objects, which are convoked and dissolved in a moment, according to poetic needs. At times they are determined by mere simultaneity. There are objects composed of two terms, one of visual and another of auditory character: the color of the rising sun and the faraway cry of a bird. There are objects of many terms: the sun and the water on a swimmer's chest, the vague tremulous rose color we see with our eyes closed, the sensation of being carried along by a river and also by sleep. These second-degree objects can be combined with others; through the use of certain abbreviations, the process is practically infinite. There are famous poems made up of one enormous word. This word forms a poetic object created by the author. The fact that no one believes in the reality of nouns paradoxically causes their number to be unending. The languages of Tlön's northern hemisphere contain all the nouns of the Indo-European languages—and many others as well.[45]

We think of this approach as nonnaturalist world-building. It does not try to be realistic but instead follows an idea and sees where it leads, consistent only with its own internal logic. It opens our eyes to another way of considering an object's place in the world in relation to our senses, perception, and language—more related to ideas, different systems of reality, and how we understand reality itself. Can a world like this ever be designed or is it destined to only ever be made from words?

For us, maybe, impossible colors are the colors of ideas lying just beyond our intellectual reach, still unknown, unthought, and unimagined. Yet once they become visible to the collective imagination, they color the world in entirely new ways.

Dunne & Raby, *A Pocket Universe in the Home*, 2023. Computer-generated image: Carolyn Kirschner.

Having read a text by a theoretical physicist, Leonard Susskind, suggesting that for all intents and purposes, the space inside a magnetic resonance imaging (MRI) scanner could be viewed as a modest alternative mini universe, which we were excited about, we decided to ask one of the scientists we met during a short visit to CERN if this was possible. Their reply: this might be overstating things a little, and the space created by an MRI just has different conditions—mainly a stronger magnetic field.[46] This is what gives the machine its functionality. The earth's magnetic field is approximately 0.00006 tesla, and the magnetic field produced by an MRI scanner is 1.5 or 3 tesla, 60,000 times stronger than the earth's. Usually the hydrogen protons, which are like tiny magnets inside hydrogen atoms in the water of our body, spin in a random manner. In an MRI machine, these spinning hydrogen protons align their axes with that of the machine, some "up" and some "down." Based on the laws of quantum physics, there will always be more that point up. This difference is what is measured by the scanner. It is only when we visualize the hydrogen photons all lined up with the field of the machine that we catch a glimpse of how our own magnetic field might be differently organized in an alternative space.

Although the original MRI analogy is probably a little fanciful, we were happy to learn that there are physicists thinking about how universes could be created in labs—a field called cosmogenesis.[47] Something, that despite being far, far away from present possibilities has already found its way into fiction. Jonathan Lethem's *As She Climbed across the Table* (2011), is a romance between a female particle physicist and "Lack," a mini black hole created in a lab exploring the origins of the universe.[48]

In fact, a tiny black hole might be the first step. It was recently announced that a team of physicists built a small-scale wormhole inside a quantum computer. It is based on the idea that when you have two black holes in a state of entanglement, a wormhole will connect them. That gravitational wormholes, which are predicted by relativity, and quantum entanglement are equivalent, just described differently. Its significance is that finally, "gravity can be described in the language of quantum physics," bringing two seemingly incompatible theories together in the form of "quantum gravity." The physicists set out to "design a quantum circuit that's mathematically equivalent to a wormhole" and were able to send a qubit through the wormhole. But as skeptics have pointed out, this is a simulation rather than an actual wormhole.[49]

So is a mini wormhole an impossible object?

For us, this is interesting not just as a starting point from which to build a world, or explore the ethics of such an experiment or even what the world might look like once ideas like this become possible, but also for the impact this has on the way we think about and understand the malleability of reality. It is more about how ideas like this reveal a side to science that works on the imagination by expanding it, enlarging reality and what the world is, more like philosophy. Real, only at the edges of theoretical physics; so impossible now, but maybe one day, mini universes will be a possibility. From that new vantage point, what vistas will be begin to appear lying just beyond an imaginative horizon even further away than our present one?

In a section called "Back to the Future" in his book *Impossibility: The Limits of Science and the Science of Limits*, John D. Barrow suggests four possible futures for science based on two aspects: "whether or not there is an unlimited store of fundamental information about Nature to be uncovered, and whether or not our capabilities are limited or not."[50]

Type 1 future: nature unlimited and human capability unlimited
Type 2 future: nature unlimited and human capability limited
Type 3 future: nature limited and human capability unlimited
Type 4 future: nature limited and human capability limited

This might seem a little abstract, but it is an intriguing space to think about possibilities for world-building. Rather than looking at economic, technological, or social factors, these four concepts could be used to set out a different kind of space of possibilities. Although it is not the aim here, this would serve as a wonderful starting point to begin to imagine what each of these worlds would be like. Particularly type 3: nature limited and human capability unlimited. In such a society, where would all the intellectual energy and curiosity currently channeled into scientific discovery be redirected?

Dunne & Raby, *Objects Undergoing Space-Time Collapse*, 2023. Computer model by Lila Feldman. Rendering: Carolyn Kirschner.

"We don't want to tinker with the Ding an sich and discover that we've somehow messed up the difference between possible and impossible in our world. It's really hard to see how that could end well."[51] This slightly panicked utterance by a character in Adam Roberts's novel *The Thing Itself* is based on Immanuel Kant's idea that something might exist beyond or outside human space-time—the thing-in-itself.[52] And that any encounter with an unknowable "thing itself" might quickly lead to madness. As an artificial intelligence remarks elsewhere in the story, "But there was always this suspicion that exposing a human being, even only partially, to the unmediated thing itself would have deleterious effects. Your species is very finely calibrated not only to exist within a structuring consciousness of space and time, but to exist within very specific tolerances of those two things."[53] Our brains would be scrambled trying to process sensory input we are unable to deal with or make sense of. An idea explored further by philosopher and cultural critic Steven Shaviro in his book *Extreme Fabulations* when he discusses Charles L. Harness's 1950 short story *The New Reality* about a scientific experiment that threatens to "destroy the Einsteinian universe." "Adam Prentiss Rogers, the story's protagonist, is an 'ontologist' working for the International Bureau of the Censor. His mission is to 'keep reality as is' by suppressing any scientific research that might 'alter the shape of that reality.'"[54]

Reality here is very much reality as described by human science, and specifically space-time, suggesting the possibility that rather than being a container that every form of life exists within, space-time is in fact a product of being human. That something else lies outside it, something profoundly inhuman.

But what would this even look like? Can we make or find an object that somehow begins to suggest this impossibility. Evoking another space-time while sitting firmly within this one? That tries to suggest something beyond the human that is still legible to humans. Probably not. But like many impossibilities, it is the attempt to grasp or visualize them that is interesting.

There have been many attempts to represent extreme transformations of space-time in film, for instance, but they always seem a little melodramatic, involving lots of slowed-down stars expressed as blurry streaks, accompanied by psychedelic light effects suggesting that the collapse of space-time is akin to an acid trip. Perhaps this too is an expression of the limits of the human imagination and it is necessary therefore, to turn to imagery generated by nonhumans.

On first encountering the glitchy images produced by Apple iOS Maps' 3D flyover view that began to surface in 2012, we started to wonder if these could serve as hints of what the fruits of an artificial imagination might produce. An interview in *Wired* in 2013 with Peder Norrby, a graphics software engineer who put a set of these images together on Flickr, focused on the beauty of these "algorithmically-authored oddities," but they are more than this.[55] To us, they suggest landscapes generated by an artificial intelligence program based on Meinong's theory of objects.[56]

These images portray a sort of familiar world. Many elements are recognizable as roads, different kinds of transport, buildings, infrastructure, and landscapes, but transformed and fused in visually beautiful and structurally illogical ways. They offer a refreshing alternative to the plethora of moody, painterly, images generated by machine learning, going beyond the surface to suggest spatial and topographic glitches, as if a machine was trying to show other machines the world we humans live in, but not quite understanding the operational or even functional logic underlying our world.

The impossible infrastructure they portray suggests a version of our reality where the laws of physics have suddenly collapsed or become scrambled due to some kind of physics apocalypse or reality crash. Artists and designers sometimes use physics engines in 3D modeling packages to construct worlds with slightly modified laws of physics, such as the strength of gravity, to create what are sometimes called "nonnormal worlds," "nonclassical worlds," or even "impossible worlds." But these are nearly always adjustments to the existing world.

Peder Norrby, *Mapglitch: Digital Artefacts*, 2013. Snaps of glitches in iOS Maps in 3D mode.

In contrast, the glitchy map images we are discussing here look as though the very fabric of space-time has been torn asunder, shredded like the gravity traps in the Strugatsky brothers' *Roadside Picnic*, hinting at a nonhuman-centered form of world-building. A world where the very nature of space-time is different, where light behaves differently, and every object contained within it reflects this. As Barrow reminds us in *Impossibility*,

> Our world is governed by relativity because the speed of light is finite. We do not know why the speed of light takes the specific value that it does in our Universe. If it were much smaller, then more slowly moving objects would suffer the distortions of space and time that arise as the speed of light was approached; less energy would be available when matter was annihilated in nuclear reactions; light would interact more strongly with matter; and matter would be less stable.[57]

One of the most enjoyable things for us when reading science fiction are the glimpses provided of alternative worlds and their systems of reality. But after a while, the endless permutations of existing political formats wear thin. These images in contrast invite the viewer to imagine radically different realities based on imagined, nonhuman metaphysics.

Object 7: A Stone Raft

Dunne & Raby, *A Stone Raft*, 2023. Computer-generated image: Carolyn Kirschner.

As Mark Twain remarked in *Following the Equator*, "Truth is stranger than fiction, but it is because Fiction is obliged to stick to possibilities; Truth isn't."[58]

When an idea or scenario is presented as a future, there is a natural impulse for people to think about how to get from here to there, triggering all sorts of pragmatic and limited conversations. It is evaluated against plausibility. Or it is taken literally as a wish, desire, prediction, or prescription to be imposed by one group of people on another. At a time when "futures" have become the dominant mode of framing the "not here, not now," at least in design, magic realism might allow us to move beyond the limitations that condemn designers, including us, to forever reimagining variations on a broken reality. It can reveal pathways that lead beyond the projection of objective realities grounded in science and technology to a far larger, richer landscape influenced by literature, philosophy, and art. Unconstrained by "technological reason," it offers something a little more poetic, which if you are trying to prompt new thinking rather than provide options, seems like a good direction to explore.

It also gently signals that a project is not "real," avoiding the kind of confusion that arises when the unreal is presented as real. With magic realism, the unreal is only ever presented as unreal, whereas fake reality is unreality pretending to be real. In design especially, viewers often take the potential of something to be real for actual reality. Some designers declare this, while others hide it.

With magic realism, there is no need to leave this world or travel to the edges of the universe in search of novel ideas. In this form of world-building, small but significant adjustments are made to the world we already live in. Often unexplained and free from scientific justification, they celebrate and highlight boundaries between the real and the unreal. Things happen that could never happen in our reality, but nonetheless people carry on, only mildly disturbed. In José Saramago's *The Stone Raft* (1986), the Iberian Peninsula breaks away from the continent of Europe and begins to slowly drift across the Atlantic Ocean, becoming the setting for a political fable.[59]

As designers, could we embrace and make use of strategies like this? Such magical realist worlds might be impossible, but they are not pointless. They can serve as a stage or setting for helpfully decontextualized alternative realities.

In Saramago's *Stone Raft*, life carries on. The characters' reactions tothe event are pragmatic; everything else in the world in which the apparently impossible thing has happened appears normal or as we would expect. The result feels absurd. Magic realism usually makes no attempt to explain or justify the anomaly behind the magical event. Its justification lies in the conceptual possibilities it allows for in the narrative, pleasure it provides, and feeling of strangeness that comes from a familiar world being tweaked. Bringing magic realism to design has always felt a bit uncomfortable, almost too easy, but we're not so sure anymore.

This tension between rational and ungrounded speculation in relation to literature is something science fiction theorist Darko Suvin has grappled with in several of his writings. For Suvin, science fiction is "a literary genre whose necessary and sufficient conditions are the presence and interaction of estrangement and cognition, and whose main formal device is an imaginative framework alternative to the author's empirical environment."[60] He suggests there is a spectrum with naturalistic worlds at one end–that is, worlds that resemble those of the author and reader–and at the other, invented worlds–what he terms "novum"–that generate defamiliarization, or what he calls "cognitive estrangement." This is what separates critical science fiction from fantasy, which may be inventive and pleasing, but lacks a critical dimension:

> The *folktale* also doubts the laws of the author's empirical world, but it escapes out of its horizons and into a closed collateral world indifferent to cognitive possibilities. It does not use imagination as a means of understanding the tendencies latent in reality, but as an end sufficient unto itself and cut off from the real contingencies...It simply posits another world beside yours where some carpets do, magically, fly, and some paupers do, magically, become princes, and into which you cross purely by an act of faith and fancy. Anything is possible in a folktale, because a folktale is manifestly impossible.[61]

Although later, in the 2000s, reflecting on how times had changed since the 1970s, he begins to question this: "Let me therefore revoke, probably to general regret, my blanket rejection of fantastic fiction. The divide between cognitive (pleasantly useful) and non-cognitive (useless) does not run between SF and fantastic fiction but inside each–though in rather different ways and in different proportions, for there are more obstacles to liberating cognition in the latter."[62]

The Stone Raft can be viewed as a stage or setting rather than a scenario. It provides a framework that estranges on a larger scale. Is it finally time to look to magic realism as a way of moving beyond the coziness of futures thinking in design, produced by following extrapolative threads tied to existing reality? This is something that has long been explored by artists and designers interested in Afrofuturism and its many variations, but for those of us trapped within a tradition of scientifically grounded speculation like science fiction, this leap is more difficult to make.

Magic realism could offer pathways forward that leave behind the problem of reproducing existing mindsets through material and visual reification. Allowing us instead to constructively undermine them, creating room for new possibilities and imaginaries to emerge. Once we leave behind the overly rational that defines much of Western speculative fiction in design, many other forms of speculation begin to emerge. Moving from questions of "How do we get from here to there?" or "Is it plausible?" to ones of "Is it interesting, and what new thoughts does it make possible?"

Maybe as designers, we could embrace a remark Borges made during a lecture on poetry he gave at Harvard in the 1960s. He claimed that no argument ever convinces, as we immediately respond with a counterargument that challenges, unpicks, and tests it. Whereas something hinted at, such as through metaphor, engages us differently and enters the imagination more easily. Not to convince as such, but to exist within the mind.[63]

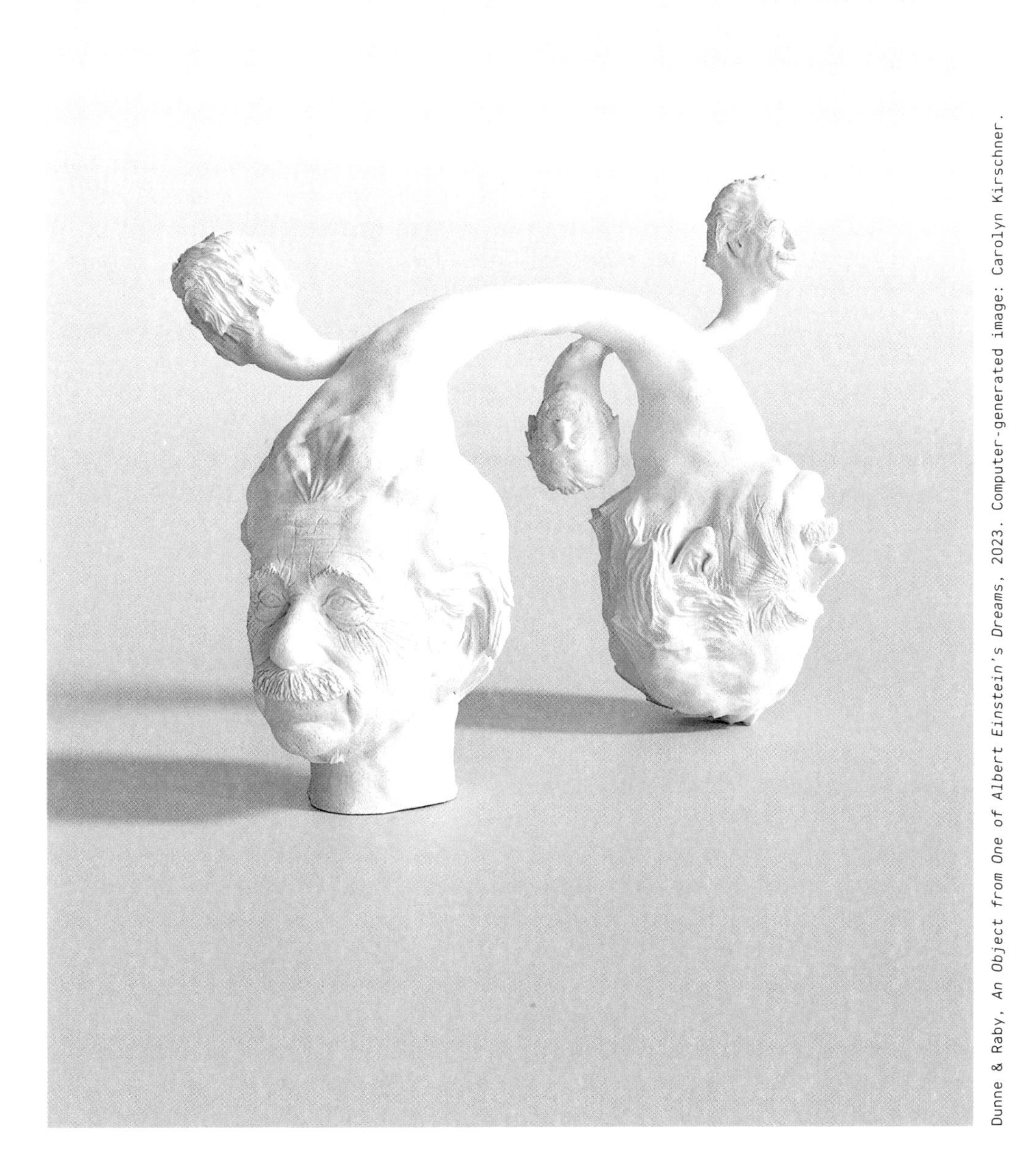

Dunne & Raby, *An Object from One of Albert Einstein's Dreams*, 2023. Computer-generated image: Carolyn Kirschner.

In Alan Lightman's book *Einstein's Dreams* (1992), the author presents thirty scenarios for different concepts of time based on Einstein's theory of relativity.[64] Each one is expressed in the form of a vignette from an imagined dream Einstein might have had while working in a patent office in the Swiss village of Berne during 1905. Labeled only with a date, they describe several simple, everyday actions or interactions that reveal the implications of these different temporalities. Some examples:

April 19, 1905:	a world where time has three dimensions
April 23, 1905:	a world with two times–mechanical time and bodily time
May 3, 1905:	a world where cause and effect are erratic
May 8, 1905:	the date the world will end is known
May 10, 1905:	a world where the texture of time is sticky
May 11, 1905:	the passage of time brings increasing order
May 14, 1905:	a place where time stands still
May 20, 1905:	a world without memory
May 29, 1905:	everything is in motion in order to slow time down
June 2, 1905:	time flows backward
June 5, 1905:	a world where time is a sense
June 10, 1905:	time is a quality not a quantity (time exists but cannot be measured)
June 11, 1905:	a world without a future (no one can imagine the future)
June 15, 1905:	time is a visible dimension
June 20, 1905:	time is a local phenomenon
June 22, 1905:	a world where the future is fixed
June 25, 1905:	a world of countless copies

What we like about this is how on first reading, without a mention of Einstein, one might think this is a book based on magic realism. One factor is changed–in this case, the concept of time–and everything else remains as one would expect. Within each alternative time paradigm, people continue to act rationally. It is an approach that encourages the imaginative freedom of magic realism but is grounded in science. In some ways it is still magical, as each concept of time is an exaggeration that in reality has only minor implications at the human scale.[65] It follows a tradition in science writing that dramatizes potentially challenging concepts such as relativity or quantum mechanics by applying their principles at the human scale using macro objects. Theoretical physicist George Gamow's *Mr Tompkins in Paperback* is probably one of the best-known examples.[66] Each chapter is based on a dream Mr. Tompkins has while falling asleep during various physics lectures. In one scene, the speed of light is slowed down to fifteen kilometers per hour to illustrate effects such as objects becoming shorter as the viewer travels faster. There is also physicist Robert Gilmore's *Alice in Quantum Land: An Allegory of Quantum Physics*, where Alice visits an amusement park smaller than an atom and each attraction illustrates some aspect of quantum mechanics.[67]

Besides time-traveling machines and portals to wormholes, designed objects are rare in chronologically shifted worlds. When they do appear, they are essentially versions of watches or clocks. Although intellectually and conceptually fascinating, it is extremely difficult to develop a design project based on time. We have used *Einstein's Dreams* as a brief for a number of workshops where participants were asked to select one of the worlds from the book and design a timepiece or other object for it. It's a surprisingly difficult thing to do.[68] While some people added another world to the list, and others illustrated the experience of living in different time systems through a performance or scene, few, if any, designed a new object for one of the worlds, beyond a variation on a clock, that could coexist in our world.

In one of our Designed Realities classes, designer Kalyani Tupkary drew on her cultural heritage to unpick Western ways of organizing time by developing a series of calendars that conceptualized time differently called the Calendar Collective.[69] Each calendar was accompanied by a voice message of someone arranging to meet using that calendar. For example, one that fluctuates with the phases of the moon. Some nights are short while others long. One where the day is as large as the sunlight one gets. One where a day is twenty-four-hours long, but some hours are more colorful than others. One where time blooms and withers with the opening and closing of flowers. And one where a day is either an "even" or "odd." Evens are born on even days. They work on even days and rest on odd days. Odds do the opposite. But evens get an additional holiday at the year's end. When she began the project just before COVID-19, many of her ideas were met with bemusement when exposed to audiences outside the class. A few months later that had all changed, and people could see that how we relate to time could be very different. Many of us who experienced various forms of lockdown also saw how our relationship to time changed in strange and unexpected ways.

Designing for time seems like an almost impossible endeavor better suited to literature or cinema. But we'd like to mention one more example. Impossible from our Western perspective, but potentially real in another cultural setting. A concept we first encountered in the title of the South Korean contribution to the 2015 Venice Biennial by artist duo Moon Kyungwon and Jeon Joonho called *Chukjibeop* and *Bihaengsul*.[70] A North Korean concept that translates roughly into "The Ways of Folding Space & Flying," a "hypothetical method of folding space and of allowing one to travel a substantial distance in a short space of time."[71] It has been claimed that Kim Jong Un can do this, although the *Rodong Sinmun* newspaper, which serves as the official newspaper of the Central Committee of the Workers' Party of Korea, has disputed this.

And it is here that we find an object for this section of the archive.

Dunne & Raby, *An Object from an Alternate Quantum Imaginary*, 2023. Computer-generated image: Carolyn Kirschner.

One of the most enjoyable aspects of giving talks outside the United States and Europe is when the questions move from universal to more culturally specific approaches to speculation, imagination, dreamworlds, fiction, and metaphysics along with their relationship to everyday life and design. These are too often suppressed in an effort to embrace a more international approach usually influenced by a Hollywood style of design futures. Although this is beginning to change, and over the last few years genres such as Afrofuturism have now firmly entered the design mainstream, with many others such as Sinofuturism, Yugofuturism, Hungarofuturism, and Baltic ethnofuturism bubbling just beneath the surface.

As Frédéric Neyrat writes of Afrofuturism in *The Black Angel of History, Afrofuturism's Cosmic Techniques*, "Here we can see the stark difference from the geo-technology of the Anthropocene, which, faithful to the modern categorical imperative, postulates that nothing is impossible, that everything can and even must be produced, realized, from nuclear power to fracking via geoengineering and GMOs (genetically modified organisms). By contrast, Afrofuturism develops a technique of the impossible: the revelation of the impossible at the very heart of technological possibility."[72] Something that is clear in the work of writers like Octavia Butler, who has been imagining beautifully alternative worlds for decades, such as in the Xenogenesis series, her collection of three stories exploring the emergence of a human-alien hybrid species with a radically different *umwelt* from humans and an ability to manipulate biology while repopulating an earth devastated by nuclear war.

Sierra Leonean artist Abu Bakarr Mansaray's drawing *One of the African Black Magic. The Witch Plane* (2008) is particularly interesting in this respect. It shows a device that looks like an armed vehicle that integrates motifs from black magic with more familiar-looking engineering components. As the artist says about his approach, "I like doing strange drawings and also designing complicated machines based on scientific ideas that sometimes exceed human imagination."[73] This idea of exceeding human imagination is particularly relevant in the context of impossible objects.

Are these the first signs of philosopher of technology Yuk Hui's notion of cosmotechnics beginning to emerge? Hui asks, "If one admits that there are multiple natures, is it possible to think of multiple technics, which are different from each other not simply functionally and aesthetically, but also ontologically and cosmologically?"[74] And suggests "that instead of taking for granted an anthropologically universal concept of technics, one should conceive a multiplicity of technics, characterized by different dynamics between the cosmic, the moral and the technical."[75]

China's quantum satellite Miscius, launched in 2016, has demonstrated a secure method of quantum messaging between two ground stations more than a thousand kilometers apart. It was the first dedicated quantum communications satellite to make use of what is called "entanglement-based quantum-key distribution."[76] The purpose of experiments like this is to bring quantum cryptology closer to reality, and with it, the dream of unhackable communications or a quantum internet.[77] Whereas many countries are neck and neck in the race to develop the first quantum computer, China is widely acknowledged as a leader when it comes to entanglement-based quantum communication.[78] Although Miscius does not make use of forces any stranger than the already weird "spooky action at a distance" of quantum entanglement, the imagined world an object like this belongs to aligns with a different kind of quantum imaginary than that suggested by quantum computing.

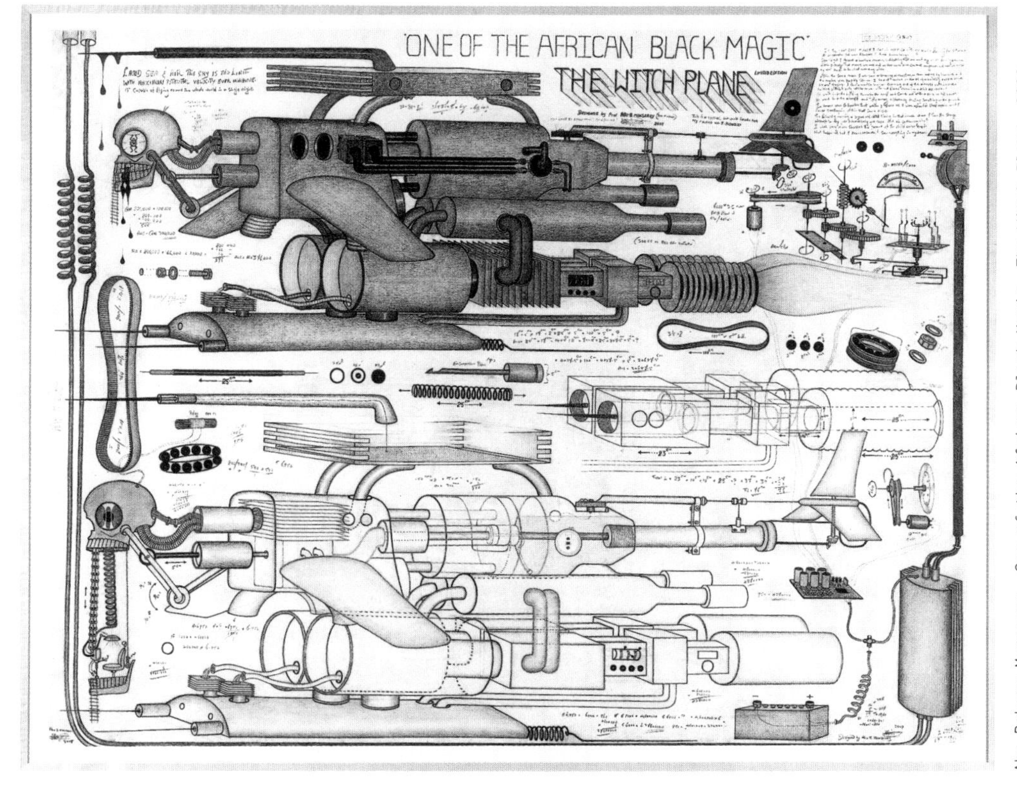

Abu Bakarr Mansaray, One of the African Black Magic. The Witch Plane, 2008. Ballpoint, colored pencils and graphite on paper, 150 × 200 cm. © Abu Bakarr Mansaray. Photo: Maurice Aeschimann. Courtesy of the Jean Pigozzi African Art Collection.

Dunne & Raby, *A Human Imagined through a Generalized Nonhuman Umwelt*, 2023. Computer model by Franco Chen. Computer-generated image: Carolyn Kirschner.

I think it would be useful if the concept of the umwelt were embedded in the public lexicon. It neatly captures the idea of limited knowledge, of unobtainable information, and of unimagined possibilities. Consider the criticisms of policy, the assertions of dogma, the declarations of fact that you hear every day—and just imagine if all of these could be infused with the proper intellectual humility that comes from appreciating the amount unseen.

—David Eagleman[79]

While reading McKenzie Wark's essay "Paradoxical Modernismo[-9088a" in *Futurity Report*, we were taken by a reference to J. B. S. Haldane's essay "Possible Worlds," so we dug a little deeper.[80] In it, he develops several playfully stimulating speculations on what various animal philosophies or systems of thought might be like based on their senses. As biologist Jakob von Uexküll writes in the opening of his 1934 essay "A Stroll through the Worlds of Animals and Men: A Picture Book of Invisible Worlds," "The first task of *Umwelt* research is to identify each animal's perceptual cues among all the stimuli in its environment and to build up the animal's specific world with them."[81] What he calls the phenomenal world or self-world of the animal. The world as it appears to an animal based on its sensory uniqueness. A world where the invisible becomes visible, where discreet entities fuse into singular ones, and so on. Haldane notes, "Remember that to a dog a thing's smell is its most real quality."[82]

Von Uexküll calls the world each creature constructs through its unique sensory organs an "umwelt." The world of a tick, for example, is built around the detection of butyric acid, which indicates the nearby presence of a mammal, triggering it to drop from a branch or blade of grass and attach itself to the animal or human passerby. It can wait for years for this happen. Time and space is sensed differently in the nonhuman world.

For von Uexküll, this realization is not trivial: "We are easily deluded into assuming that the relationship between a foreign subject and the objects in his world exists on the same spatial and temporal plane as our own relations with the objects in our human world. This fallacy is fed by a belief in the existence of a single world, into which all living creatures are pigeonholed. This gives rise to the widespread conviction that there is only one space and one time for all living things."[83] Increasingly, designers are beginning to embrace the idea that there is not just one world, the human world, but instead many. And that each creature or life-form has its own world, just as real as ours. All melding and fusing into worlds within worlds. When we enter this complex world of worlds, we no longer retain our human shape or are made from human stuff. We undergo a figurative, material, and conceptual transformation. We take on new forms, new materialities, and new meanings. Our presence in these worlds is radically different. Maybe even monstrous to a nonhuman.

There are many projects that attempt to show us what the human world looks like to another creature—a bat, for example—or what it sounds like to an elephant, a sort of sensory cut and paste. But as Thomas Nagel points out in his famous paper What Is It Like to Be a Bat?" it is impossible for us to ever know what it is like to be another animal. We can never experience their world:

> Our own experience provides the basic material for our imagination, whose range is therefore limited. It will not help to try to imagine that one has webbing on one's arms, which enables one to fly around at dusk and dawn catching insects in one's mouth; that one has very poor vision, and perceives the surrounding world by a system of reflected high-frequency sound signals; and that one spends the day hanging upside down by one's feet in an attic. In so far as I can imagine this (which is not very far), it tells me only what it would be like for me to behave as a bat behaves. But that is not the question. I want to know what it is like for a bat to be a bat. Yet if I try to imagine this, I am restricted to the resources of my own mind, and those resources are inadequate to the task. I cannot perform it either by imagining additions to my present experience, or by imagining segments gradually subtracted from it, or by imagining some combination of additions, subtractions, and modifications.[84]

In other words, we can never know.

But for us, even if impossible, it is still interesting to speculate on how we might appear differently, within nonhuman umwelten. Not in a scientifically accurate way or for a particular species, but through the lens of a sort of imagined, generalized, nonhuman umwelt that reflects us back to ourselves, constituted differently. Encouraging a shift in mindset that requires us not to become other but rather to unbecome human. To undo the human (form) and begin to imagine how we might be present in a rich and complex world of many worlds.

The point is not to accurately represent these worlds but instead to acknowledge that the human umwelt is just one of many. That there are other ways of constituting the world that might appear alien to us. Perhaps this is the impossible object in this section—a representation of a new human figure, riddled with conceptual contradictions. Based as it is on human speculation about how a human might *appear* or be *present* to a nonhuman umwelt, yet still legible to a human. Encouraging a shift in mindset. Yet still depicted as something visual, the primary sensory channel for us humans.

Perhaps, as Eagleman suggests in the quote at the start of this section, a wider appreciation of the idea of umwelten might help us all understand that each person is limited by their own world, shaped by specific sensory apparatuses and conceptual processes, which today includes the media and other technologies through which we make sense of the world out there.

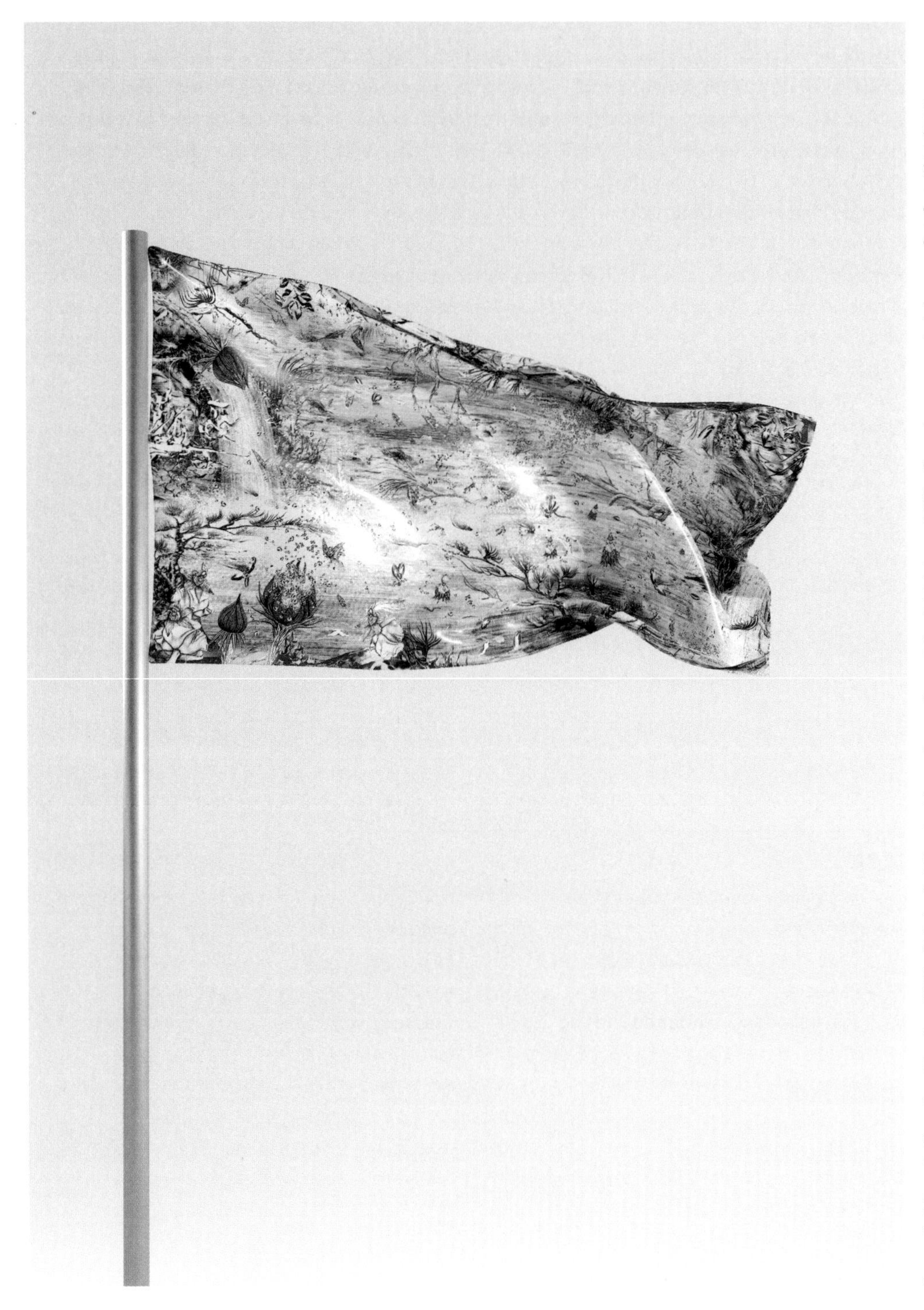

Dunne & Raby. *A Flag for Biomia*, 2023. Original illustration: Kyung-Me and David Lichen, 2020. Computer-generated image: Carolyn Kirschner.

On a trip to Beijing in 2017, one of us visited an exhibition at the National Art Museum of China showing a selection of gifts the Chinese government had received from other nations.[85] It was fascinating to see so many strange objects side by side with no connection other than they were given to the government. Each one of course had other meanings and symbolism not always apparent to a casual viewer. One couldn't help but speculate on why a particular gift was chosen by a particular country at a particular moment in history and what it could have meant. Gifts have long been a mainstay of diplomatic relations, often acting as a form of soft power promoting the values of the giver nation, or in the case of John Kerry's gift of two large Idaho potatoes to his Russian counterpart Sergei Lavrov in 2014, they can also serve other purposes.[86]

There is something appealing about working with a genre that has existed for several hundred years but still has not been embraced by design. Viewing designed objects through this lens might lead to a different kind of reading. From who was it given, to whom, and why. In preparation for a class on designing diplomatic gifts, we visited the UN building in New York City.[87] What each country gave or contributed was fascinating—"decor and furnishing." Something so domestic and down-to-earth took on a new level of significance. Little bits of national identity patched together to make a physical tapestry of values, cultures, and aesthetics. Gift giving not to another state but instead to a community of 193 states has its own complexities and pitfalls.[88]

In "Gift and Diplomacy in Seventeenth-Century Spanish Italy," Diana Carrió-Invernizzi writes about the use of gifts by Spanish diplomats in Italy during the 1600s and how the gift transitioned from old ways to what we see today:

> The main conclusions that have been reached by recent studies of gift exchange in the early modern period can be summarized as follows. There had been an evolution from a focus on the economic value of the gift above all else (although this was always of importance) towards being more concerned with the ways in which gifts were made. Thus, ingenuity and imagination might supplant financial investment. This development had taken place as a result of growing awareness that ostentation might be synonymous with bad government. Excess led to a loss of prestige, with the result that, as from 1665, international treaties tended to discourage the obligation to hand over gifts.[89]

Kyung-Me and David Lichen, *Biomia*, 2020.

If we are indeed to design for a pluriverse, then difference is something to be navigated, negotiated, and celebrated. Even if a pluriverse fails to materialize, there are clear signs that a process of deglobalization is underway as new blocs form around different technologies, materials, supply chains, and trade. Diplomatic gift design might become a necessary, even essential part of pluriversal politics. But even if not, it is still interesting to consider the diplomatic gift as a designed object that knowingly embodies other modes of being, concisely infused with a worldview alien to its receiver.

Biomia is an idea, a metaphor, for thinking about borders and mobility through the lens of a biome. It emerged from a series of workshops bringing designers, artists, historians, social scientists, international relations scholars, and others together as part of a Mellon-funded Sawyer Seminar series called Imaginative Mobilities that we co-led with Victoria Hattam

(politics), Miriam Ticktin (anthropology), and T. Alexander Aleinikoff (immigration and refugee law) at The New School from 2017 to 2019.

Although not originally intended to be a flag, could this intensely detailed image created by illustrator Kyung-Me (and David Lichen), who sat in on the sessions, to reflect the interdisciplinary discussions about biomia serve as a flag? Probably gray from a distance, but up close, teeming with life. A flag that changes depending on proximity. Most flags remain the same near or far. This one appears stained or discolored from a distance, and then up close, it reveals an intensely detailed surface. Itsignals that air, contrary to a modernist view of space as something empty, is fulsome. Teeming with life and other entities. The closer we look, the more is revealed. At odds with the immediately graspable world suggested by typical flags, possibly with the exception of Turkmenistan's post-1992 flag, which with its elaborate panel of imagery traditionally used in carpet design representing its five main Turkmen tribes, claims to be the most complex existing flag. Biomia's flag goes further. It is an unflag-like flag design, an impossible flag.

Biomia's identity lies in its irreducibility to simple, convenient images, symbols, icons, and even individuals. Echoing Merlin Sheldrake in *Entangled Life*,

> Many scientific concepts–from time to chemical bonds to genes to species–lack stable definitions but remain helpful categories to think with. From one perspective, "individual" is no different: just another category to guide human thought and behavior. Nonetheless, so much of daily life and experience–not to mention our philosophical, political, and economic systems–depends on individuals that it can be hard to stand by and watch the concept dissolve. Where does this leave "us"? What about "them"? "Me"? "Mine"? "Everyone?" "Anyone"?.[90]

Object 12: A Vegetable Lamb

Dunne & Raby, *A Vegetable Lamb*, 2023. Computer-generated image: Carolyn Kirschner.

An image in Margaret Robinson's wonderfully titled *Fictitious Beasts, a Bibliography* (1961) shows a strange plant-animal known as the Vegetable Lamb.[91] It is included for being cited in Claude Duret's *Histoire admirable des plantes etc.* (1605). For many years, rumors of this zoophyte, said to exist in Central Asia, circulated throughout the West. Some thought the lamb was the fruit of a melon-like plant, while others believed it was part of the plant and would die if separated. It was thought to grow on a stalk that was sometimes interpreted as an umbilical cord that also served as a tether allowing it to eat grass and plants around it. Once they were gone, it died. In another book, *The Vegetable Lamb of Tartary* (1887), Henry Lee writes that its meat tasted sweet like honey and that its wool was used to make clothing for those living locally. Its only predators, besides people, were wolves.[92]

This of course is but one of many impossible creatures thought to exist just beyond the boundaries of Western geography. But if we were to substitute knowledge for geography, what is today's (conceptual) Vegetable Lamb–an idea or notion that we believe in that will be shown in centuries to come to be a mere figment of our collective imagination? So wrong it is hard to believe we once thought it true. A future impossible object.

For us, the Vegetable Lamb brings to life the history of ideas, and how concepts, entities, objects, and other "things," whether material or not, can appear and disappear over time, or move from the real to the unreal, and from the actual to the imagined. Historian Darren Oldridge puts nicely:

> If rational people in the past could believe in demonically possessed apples, and execute pigs, witches and heretics, then rational people today should consider the potential "strangeness" of their own ideas. To put it another way, will people in the future find our beliefs as ridiculous as the ideas of witch-finders now seem to us? This shift in perspective has some fascinating implications. First, it draws attention to the process by which all people–including us–make sense of things around them.[93]

The Vegetable Lamb is a reminder or even emblem for the notion that reality is a metaphysical work in progress, and that the past is in many ways a far stranger and more alien world than most of us can ever imagine when attempting to think about futures.

As Bernardo Kastrup writes in *Meaning and Absurdity*, "Science does not progress through a steady refinement of a worldview, but by throwing out worldviews in favor of new, previously unthinkable ones. One cannot help but wonder which of the certainties we currently hold about the world will have to be discarded in the near future."[94]

What would design projects look like that put us into radically different relationships to what already exists or might exist? That start from different ontologies? That embrace ideas that might, in the here and now, seem absurd, odd, or unrealistic, maybe even impossible, and begin to visualize what new realities might begin to look like? Not as prescription, but as prompts for further thought and imagining.

Writing a little further on in the same text, Kastrup states, "[Thomas] Kuhn goes as far as to suggest that the very world scientists live in—in terms of their perception gestalts—changes after a paradigm transition, so that scientists actually begin to see different things. When paradigms change, the world itself changes with them."[95] Can designers, working with philosophers and other makers of ideas, weave some of this thinking into the materiality of everyday life to hint at ways of seeing the world differently? Attempting to give tentative form to what Claire Isabel Webb, writing in Noēma, calls "future epistemes, shimmering over the event horizon of knowability, that have not yet taken shape."[96]

3 Quantum
Common Se
Images, C
and Metap

nse: New
oncepts,
hors?

Chapter 3

Quantum Common Sense: New Images, Concepts, and Metaphors?

How might design connect with the abstract and complex field of quantum mechanics along with its many paradoxical, contradictory, and counterintuitive ideas and objects?

As we mentioned above, a first point of contact might be the quantum computer as it begins to enter the public imagination. Partly due to the promise of massive increases in computer power and the conveniences this will or might bring, but in relation to geopolitics too, where a quantum race is playing out. This race is not only based on increased computing speed but also uncrackable codes (quantum cryptography) and new communication media that make use of entanglement (the quantum internet). Or more idealistically, promises of facilitating phenomenally accurate predictions due to an ability to process vast amounts of data, like a sort of quantum Laplace's demon. There is even talk of quantum computing giving rise to entirely new kinds of matter with properties not found anywhere else in nature.[1] And some argue that quantum computers might be more energy efficient than classical computers for certain calculations.[2]

But this is not what interests us.

IBM Quantum scientist Dr. Maika Takita in lab. Reprint Courtesy of Connie Zhou for IBM Corporation ©.

The idea of a quantum computer rather than its applications is far more interesting. In the 1980s, physicist Richard Feynman suggested that if we are to truly model natural systems, which are effectively quantum systems, then we will need something that didn't exist at that time—something like a quantum computer. Later in the same decade, David Deutsch, another physicist and advocate for the MWI of quantum mechanics, conceived of a device, what we now think of as a quantum computer, for testing the MWI.[3] For Deutsch, MWI explains the advantage quantum computers have over classical ones. They are operating in many worlds simultaneously, and if we could just get them to self-report on what was actually going on while they computed, we would finally have proof of the MWI: "Deutsch posited an artificial-intelligence program run on a computer which could be used in a quantum-mechanics experiment as an 'observer'; the A.I. program, rather than a scientist, would be doing the problematic 'looking,' and, by means of a clever idea that Deutsch came up with, a physicist looking at the A.I. observer would see one result if Everett's theory was right, and another if the theory was wrong."[4] Although this is by no means a commonly held belief among scientists, it does hint at an intriguing fusion of philosophy, science, and everyday life that quantum computers, at least as an idea, could manifest.

With the entry of quantum computing into daily life, we might see ideas from quantum mechanics, which has been around for a hundred years, begin to impact everyday life. Maybe even challenging the Newtonian imaginary that currently shapes the "contours of the possible" along with our individual and collective imaginations, as Michael P. A. Murphy puts it in *Quantum Social Theory for Critical International Relations Theorists: Quantizing Critique*. Of course, at the scale of the objects we encounter in day-to-day life, Newtonian mechanics is completely capable of explaining what is happening. It is only when we go down to the level of atoms and subatomic particles that quantum mechanics comes into its own. But still, many of the concepts and metaphors used to explain and model everyday interactions as well as probabilities draw from a sort of Newtonian common sense. Using it as a model to explain international relations, for example, where if you apply pressure to an issue at just the right point, using just the right amount of force, then it will move off in a predictable direction; if we do this, then this will happen. But these models are struggling to explain or even make sense of the apparently paradoxical politics taking shape around us. Although in jest, this quote in the *Guardian* about the Northern Ireland Protocol and Brexit hints at what might lie ahead: "Northern Ireland would be really part of the UK customs area, but practically in the EU customs union, following European rules on tariffs and quotas. It would be simultaneously in and out—a model quickly called Schrödinger's customs union in mock homage to the physicist's theoretical cat that was simultaneously dead and alive."[5]

It is beginning to feel like our current idea of common sense is not making much sense at all and in fact seems to be at odds with the world. With the emergence of quantum computing as a potentially plausible technology likely to enter our daily lives at some point in the not too distant future, we began to wonder if this transition from laboratory to everyday life might give rise to a form of "quantum common sense" and new concepts, metaphors, and images that would allow us to better navigate as well as make sense of the new realities taking shape around us. And what that might look like.

We are not talking so much about the laws of nature; quantum theory is already expanding and modifying how we understand these. But how it might infiltrate the images, metaphors, and concepts we use to think about everyday life. As Deutsch has put it, "What it is though, philosophically, is taking a quantum world view. That is rather a revolution, but that could happen today and the only reason it has been sluggish in happening is psychological, and maybe quantum computers will help with this psychological process."[6] Not devising new metaphors to make sense of concepts from quantum theory, which feels a little too like a conceptual version of Marshall McLuhan's "horseless carriage," but the opposite. We could construct new metaphors made from quantum ideas that enable us to see beyond current restrictions and limitations. Not in relation to the workings of the physical world, but the stories we tell ourselves, and even our worldviews, in order to overcome the power of Newtonian-based ideas to structure and limit thought and action. Looking for new ways of thinking that open up new opportunities not afforded by other conceptual lenses. As Carlo Rovelli remarked in an interview in the *Guardian*, "However, I suspect that perhaps better understanding quantum theory could free us from some prejudices about the structure of reality, prejudices that obstruct us in [our] understanding of what consciousness is, whatever that word means. Quantum theory indicates that reality is more interdependent than it is in the picture classical physics paints."[7]

At its most simple, quantum mechanics, which was developed in the early twentieth century by scientists including Max Planck, Niels Bohr, Werner Heisenberg, Louis de Broglie, Albert Einstein, Erwin Schrödinger, and Max Born, among others, is a set of mathematical tools for describing how matter behaves at the atomic and subatomic levels—the microworld. For example, it cannot predict exactly where a particle will be found at a particular moment, only its probability of being somewhere at that moment. Whereas at a macro scale, if we throw an object, we can pretty much predict its behavior and where it will land based on classical or Newtonian mechanics, at the micro scale, very different laws come into play, many of them counterintuitive. Quantum mechanics allows scientists to work with these behaviors to an incredible degree of accuracy, paving the way for the development of electronics, lasers, MRI scanners, GPS systems, computers, and the many other gadgets our modern lives depend on.

Probably one of the most famous examples of the paradoxical nature of quantum theory, besides Schrödinger's cat, is the double-slit experiment where a photon, which is a quantum of light that behaves like a particle, is fired at a plate with two slits in it and recorded on a surface once it has passed through the plate. Instead of the splatter of dots one would expect to see as each photon arrived, the result is a series of

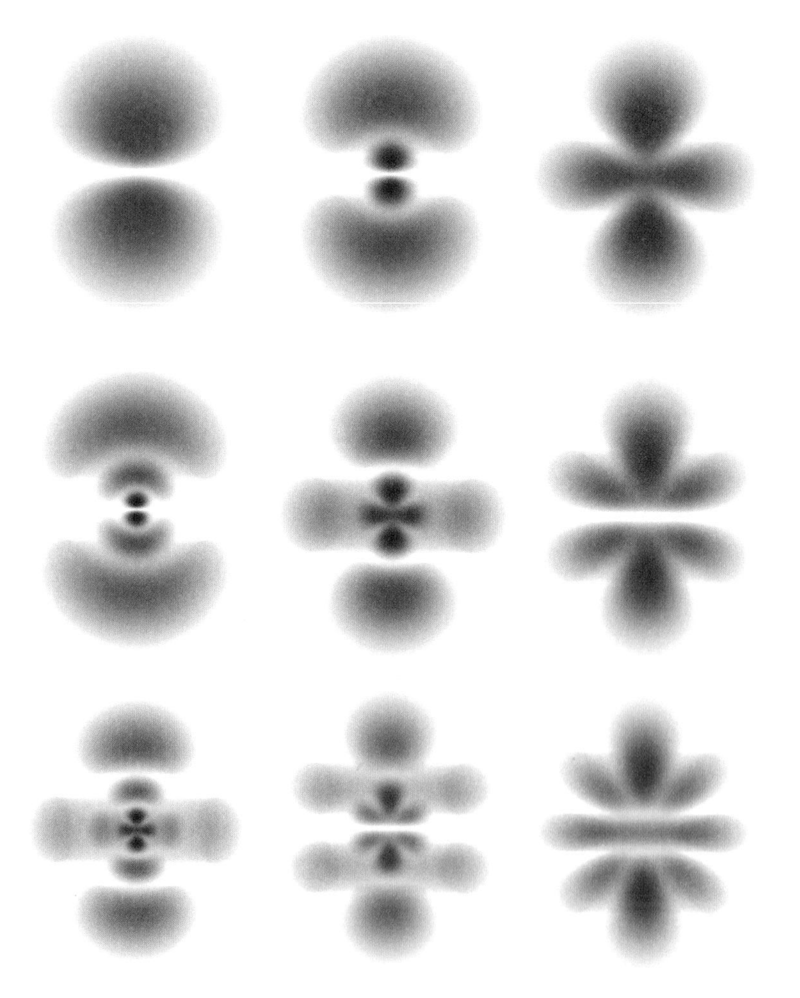

The fuzzy position of the electron relative to the proton in various stationary states of atomic hydrogen. The darker the regional orbital, the higher the probability of finding an electron in that area. Image source: Ulrich Mohrhoff, *The World According to Quantum Mechanics: Why the Laws of Physics Make Perfect Sense After All* (Hackensack, NJ: World Scientific Publishing, 2011).

interference bands that suggest light is behaving more like waves passing through both slits creating interference with each other rather than individual particles passing through one slit at a time. At this stage it simply proves what is already known: that light acts both as wave and particle. But as Philip Ball writes in an article on the experiment for *Nature*, "Odder, the pattern vanishes if we use a detector to measure which slit the particle goes through: it's truly particle-like, with no more waviness. Oddest of all, that remains true if we delay the measurement until after the particle has traversed the slits (but before it hits the screen). And if we make the measurement but then delete the result without looking at it, interference returns."[8] This suggests that it is not the measurement that makes a difference but instead *noticing* it. Which in turn raises questions about consciousness and its role in a process called "wave-function collapse," where a wave of probability—or many possible states—collapses into a single state.

Even physicists working in this area will admit that they do not fully understand what is going on here, and there are many theories that attempt to explain it. In the face of this, some scientists adopt the maxim "shut up and calculate"—a term often associated with the Copenhagen interpretation that encourages scientists to focus on doing the work rather than trying to figure out what it describes or the model of reality it suggests. But for a nonscientist, this is exactly where it gets interesting. For example, the debate on whether or not quantum mechanics actually describes reality or is a toolbox that allows scientists to carry out highly complex and accurate operations on reality. Because there is no agreement on its relationship to the actual world, there are a number of different interpretations that attempt to explain what is going on at the scale of quantum reality. A favorite among filmmakers and writers is Hugh Everett's MWI, which suggests a new branch of reality is created each time a measurement is taken or observation is made. For instance, when the box is opened in Schrödinger's famous thought experiment, if the cat is alive in the observer's universe, another reality will split off where the cat is dead—one that is entirely inaccessible and unknowable to the observer. The Copenhagen interpretation, the most commonly adopted interpretation, developed by Heisenberg and Bohr, posits quantum mechanics does not describe reality at all but rather simply offers amazingly effective tools for working with it. There are also "hidden variable" theories that assume quantum mechanics is an incomplete description of reality that will only make sense when currently unknown variables are discovered. But even if one of these interpretations eventually proves to be "correct," it is not clear what its relevance to the macro scale is, and scientists still struggle to define the boundary between the worlds of quantum and everyday objects.

As Ball points out in *Beyond Weird*,

> The physicist Leonard Susskind is not exaggerating when he says that "in accepting quantum mechanics, we are buying into a view of reality that is radically different from the classical view." Note that: a different view of reality, not a different kind of physics. If different physics is "all" you want, you can look (say) to Einstein's theories of special and general relativity, in which motion and gravity slow time and bend space. That's not easy to imagine, but I reckon you can do it. You just need to imagine time passing more slowly, distances contracting: distortions of your grid references. You can put those ideas into words. In quantum theory, words are blunt tools. We give names to things and processes, but those are just labels for concepts that cannot be properly, accurately expressed in any terms but their own.[9]

This is also why it is so interesting for artists, writers, and others intrigued by the nature of reality. We became curious about how we might approach this topic as designers concerned with material culture and everyday life. Of course, there are people who have been thinking about and exploring this in their work for decades. Especially writers and filmmakers. And several tropes have already emerged; not all are suitable for design, but by looking a little more closely at what already exists, new possibilities might be revealed.

Quantum Tropes

The most famous and for many people most intriguing trope is based on Everett's MWI mentioned above. It consists of fairly classic storylines happening within and across many worlds, generating multiple versions of each character that give rise to novel situations and tensions. For example, Frederik Pohl's *The Coming of the Quantum Cats* (1986), where the people of one alternative world or reality are preparing to invade the others. The story revolves around a limited set of characters who have very different lives in each of the worlds, possibly hinting at the complexity of forces shaping who we are and whether, or how, we fulfill our potential. In *Anxiety Is the Dizziness of Freedom* (2019), Ted Chiang adopts a more nuanced approach that moves beyond using many worlds as a backdrop to thinking about the nature of individual identity and destiny through the lens of multiples selves.[10] The story is set in a world where people can access other versions of themselves in different branches of reality known as paraselves, using a device called a prism. Here, quantum theory provides a unique conceptual lens through which to revisit ideas such as chance, individual freedom, fate, identity, and their interactions, both actual and imagined:

> Prisms had an enormous impact on the public imagination; even people who never used prisms found themselves thinking about the enormous role that contingency played in their lives. Some people experienced identity crises, feeling that their sense of self was undermined by the countless parallel versions of themselves. A few bought multiple prisms and tried to keep all their parallel selves in sync, forcing everyone to maintain the same course even as their respective branches diverged. This proved to be unworkable in the long term, but proponents of this practice simply bought more prisms and repeated their efforts with a new set of parallel selves, arguing that any attempt to reduce their dispersal was worthwhile.[11]

Another more recent variation may be because we are living in the time of artificial intelligence, predictive technologies, and with the first applications for quantum computing possibly just over the horizon, hyperdeterminism. This is the focus of Alex Garland's eight-part series *Devs*. In this story, a quantum computer with an almost unimaginable number of qubits is able to reconstruct moments in history that suggest it can also be used to predict the future. Its main protagonist is strongly opposed to the many worlds approach and more interested in his own future rather than what might have been or could be in other versions of his time line. Here, it is about determinism versus free will and what it means to be someone living in a world where it has become clear that free will might not exist. In cinema, besides serving as a narrative setting, the many worlds approach or idea of a multiverse allows franchises like the Marvel Cinematic Universe to bring back characters who have previously died in alternative time lines, giving rise to a new kind of business model.

But what about design?

There are several designers who have grappled with this. One example is Anab Jain and Jon Ardern's *The 5th Dimensional Camera* (2010), a metaphoric representation of quantum computation in the form of a fictional device capable of capturing glimpses of parallel universes. *The 5th Dimensional Camera* attempts to represent ideas from Everett's MWI of quantum mechanics. In an accompanying video setting out the science behind the project, a researcher tells the viewer that the fictional device is designed for a world where quantum mechanical concepts move from the laboratory into everyday life. The project evolved out of conversations with researchers working on quantum computing at Bristol and Oxford universities as part of a project we ran in the Design

Interactions Department at the Royal College of Art in 2010 for the Engineering and Physical Science Research Council that linked designers to sixteen research laboratories they funded at that time around the United Kingdom.

It is communicated through the stories of three characters, such as:

> Nolan is 70. He lives in a small flat by himself in north-west London. He spends his time reading, reflecting on the world outside his window and solving crosswords in the evening paper.
>
> One evening a small advert in the corner of the newspaper catches his attention. A group of scientists were looking for participants to live with a "5th Dimensional Camera" for a few months. He reaches for his phone and soon the camera is standing awkwardly in his living room.
>
> Nolan starts by taking photographs of the empty corridor outside his flat. Over time, his attention turns from reflecting on the world outside his window to a fascination with the images of the many worlds revealed outside his front door.

Anab Jain and Jon Ardern, *The 5th Dimensional Camera*, 2010.

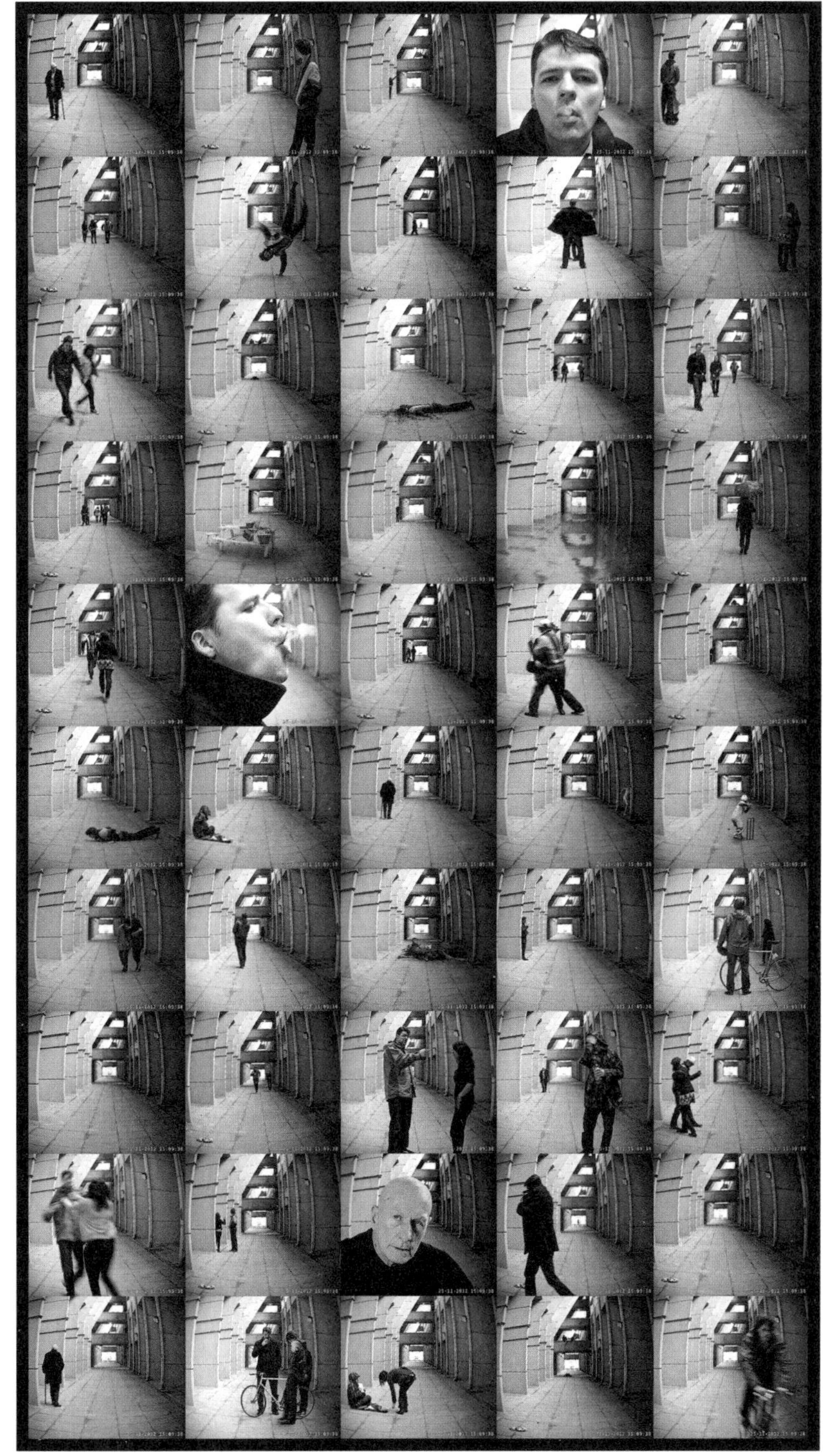

Anab Jain and Jon Ardern, *The 5th Dimensional Camera*, 2010.

The Quantum Parallelograph (2011) by Patrick Stevenson Keating (now Studio PSK) examines the scientific and philosophical ideas surrounding the theory of quantum physics and multiple universes by simulating a device that provides news about other versions of yourself from a parallel world.

The *Many-Worldian Artifact Hunter* by Shashwath Santosh done in one of our Quantum Common Sense classes looks at many-worldian emotions such as "quantum guilt," "classical arrogance," "many-worldian existentialism," "quantum narcissism," "many-worldian envy," and "quantum hyperoptimism." Inspired by a talk from physicist Sean Carroll on the MWI of quantum mechanics, the project makes use of an actual app called the Universe Splitter to produce a map of an actual route taken in Manhattan combined with a route taken in a parallel Manhattan.[12] The idea is to test the emotions produced if one truly embraced a many-worldian mindset while traveling across the city. At each junction, a choice is entered into the Universe Splitter, which immediately contacts a laboratory in Geneva, Switzerland, and then connects with a device that releases single photons into a partially silvered mirror so that each photon will have two possible paths. According to the MWI, the photons will take both paths, but in separate universes, thereby splitting the universe into the route that Shashwath took and the one he didn't. To record

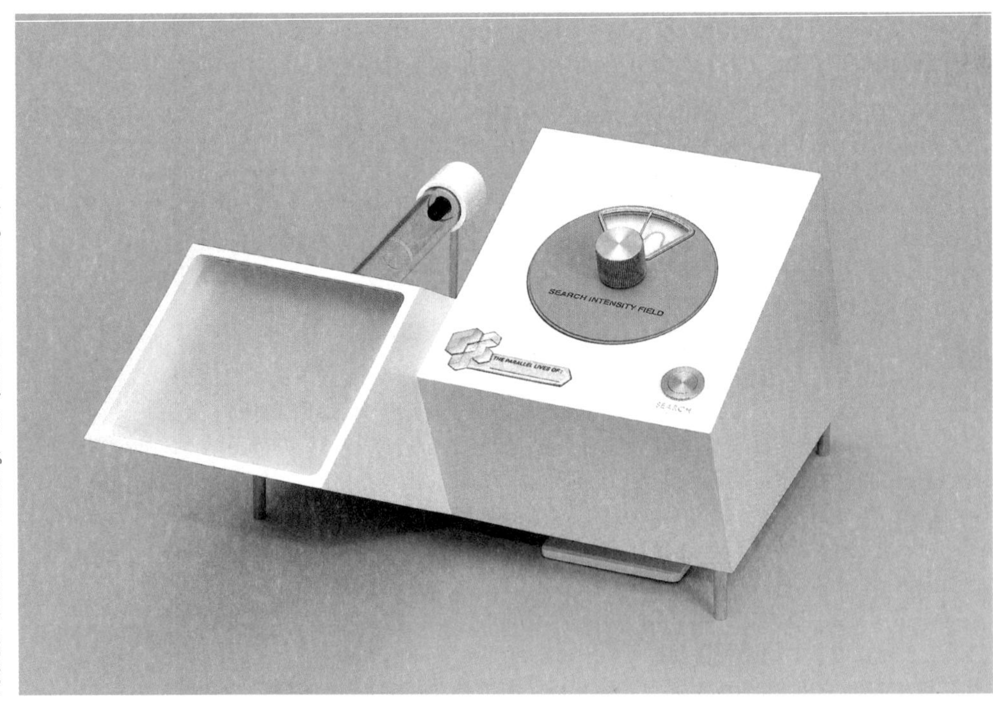

Patrick Stevenson Keating, *The Quantum Parallelograph*, 2011.

Shashwath Santosh, *Many-Worldian Artifact Hunter*, 2022.
Photo: Shashwath Santosh, Krithi Nalla.

each decision, he picked up discarded artifacts from whichever universe he found himself in, placing them in one of two bags labeled Universe A and Universe B. One could ask if a dice would do instead, but a dice is not truly random. In theory, the outcome could be predicted if everything was taken into consideration. For a many-worldian, it is not just how a decision is made but also the consequence of that decision—the creation of additional worlds and what it feels like to be implicated in the extravagant generation of new worlds, even if theoretical.

Quantum as Medium

Possibly the most challenging approach, as you need some technical abilities to do this, is working with quantum as a medium. Libby Heaney is a rare artist with a background in quantum physics who works directly with quantum computers. In *Cephalopod Aliens (Studies of Tentacular Creatures with Quantum Computers)* (2019), Heaney created a series of watercolor paintings that she then scanned and parsed through a quantum algorithm using a cloud quantum computing system. In the project description, the artist writes, “All of the frames besides the watercolor images are generated through the quantum effects of superposition and entanglement.” And “throughout the video the original cephalopod painting always exists, but due to quantum entanglement between the pixels it is delocalised across the entire picture frame, accessing gateways to other universes it reveals previously hidden perspectives of the original image, until it eventually recoheres.”[13]

One of the challenges with work like this, though, is in producing something that is uniquely quantum, despite making use of quantum processes and ideas, it does not simply look like a nonquantum work of art. For us, it is important that quantum-based art and design somehow lead to or suggest other ways the world could be, or even look, rather than arriving at something seemingly familiar by novel means. In quantum computing, the concept of “quantum advantage” describes a situation where quantum computers can solve problems classical computers cannot. Could this also be helpful in looking at art and design working with ideas from quantum mechanics?

Quantum as Lens

The word “quantum” has been used as a prefix for many topics–quantum diplomacy, quantum biology, quantum fiction, quantum computing, and so on, and we are certainly not suggesting it should prefix design. Some of these disciplinary fusions use quantum theory to develop new possibilities that traditional tools in those fields are unable to address, like the application of quantum ideas to game theory. In other fields, such as quantum biology, researchers hope quantum theory will explain partially understood and sometimes mysterious mechanisms, like the avian (quantum) compass of the European robin. Along with many other animals, it can sense the earth’s extremely weak magnetic field to aid navigation using a chemical compass that rather than distinguishing between the magnetic north and magnetic south, can distinguish between the equator and pole like a form of avian inclination compass. Current thinking indicates it makes use of quantum entanglement in ways that are still being researched.[14]

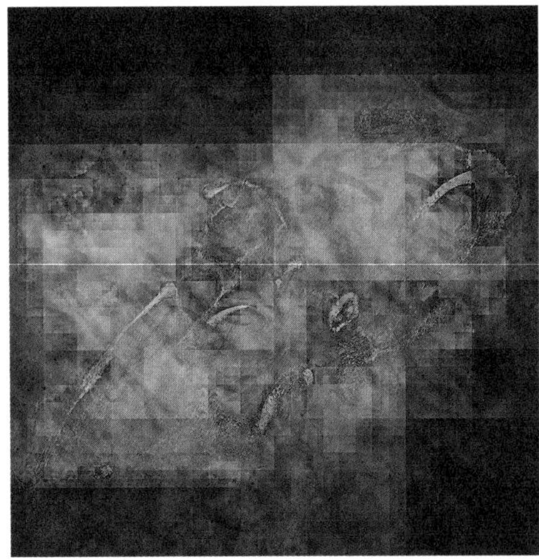

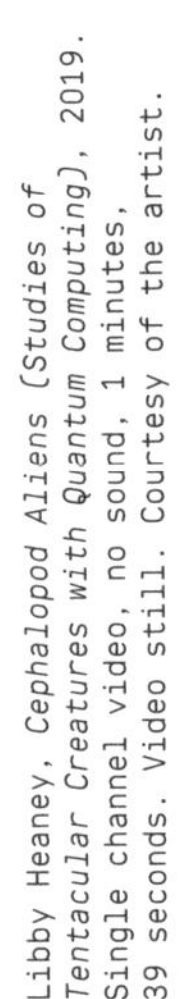

Libby Heaney, *Cephalopod Aliens (Studies of Tentacular Creatures with Quantum Computing)*, 2019. Single channel video, no sound, 1 minutes, 39 seconds. Video still. Courtesy of the artist.

Other approaches draw analogies between quantum phenomena and different fields in the hope that quantum theory might help develop new insights and perspectives in those areas.[15] This comes with many risks, however, simply providing a new vocabulary for old concepts –a sort of intellectual rebranding. Some of the most interesting crossovers have been made in the social sciences. For example, Michael P. A. Murphy has explored how ideas from quantum theory in the form of a “quantum imaginary” might lead to new thinking in the field of international relations.[16] The aim is to overcome the power of a Newtonian-based vocabulary to structure and limit thought and action. Not to celebrate weirdness for its own sake, but to free up imaginative capacity and redefine the boundary between the possible and the impossible.

Nicholas Harrington, a political philosopher and member of Project Q, an initiative run by a group of social scientists at the Centre for International Security Studies at the University of Sydney that “investigates the geopolitical and societal implications of quantum innovation in computing, communications and artificial intelligence,” has written several pieces for their website examining connections between quantum theory and political thought.[17] One of the most rigorous explorations in this field is Alex Wendt's *Quantum Mind and Social Science* (2015), which looks at the "implications for social science of the possibility that consciousness is a macroscopic quantum mechanical phenomenon—in effect, that human beings are walking wave functions."[18] Connections have also been made with ideas from Eastern thought, such as Fritjof Capra's *The Tao of Physics: An Exploration of the Parallels between Modern Physics and Eastern Mysticism.*[19] Here, quantum theory is not being used to justify mysticism but instead to highlight similarities. Maybe, as nonscientists steeped in reason, rationality, and science, quantum physics provides a bridge to ideas and concepts usually kept at a distance. A rational way into the seemingly irrational, so to speak.

But there are significant conceptual challenges to using quantum as a lens. As Karen Barad has put it so beautifully, "I am not interested in drawing analogies between particles and people, the micro and the macro, the scientific and the social, nature and culture, rather, I am interested in understanding the epistemological and ontological issues that quantum physics forces us to confront, such as the conditions for the possibility of objectivity, the nature of measurement, the nature of nature and meaning making, and the relationship between discursive practices and the material world."[20]

Map Project Office and Universal Design Studio, IBM Quantum System One, 2019. Reprint courtesy of IBM Corporation ©.

Quantum Materiality

As designers, we are always interested in the fusion of ideas and materiality. And in some ways the quantum computer is a physical point of contact with these ideas. A place where many different kinds of quantumness become physical. Understandably, engineering currently drives how quantum computers are configured, but in 2019, IBM worked with Universal Design Studio and Map to explore what a quantum computer might look like given a little more leeway. The IBM Quantum System One as it is known used specially developed glass to protect the qubits from interference and create a striking image for the computer, but it still feels a little restrained by its technology.

Even in cinema, where there is usually more freedom to examine what technology might look like, reality dominates. In *Devs*, a TV series centered on a fictional quantum computing start-up called Amaya, a great deal of effort has gone into the design of the machine itself and its architectural setting. It appears to be a slightly more elaborate version of quantum computers seen in press images from IBM and others. Intricate gold detailing, a golden chandelier-like object, all suspended in splendid isolation. But it is different, more detailed, more religious looking. Not a surprise when you discover the series title can also be read as "Deus." But it is a little too close to the state of the art. Its setting, on the

IBM Quantum System One on display at CES in 2020. Reprint courtesy of IBM Corporation ©.

other hand, is more radical; the whole room, a lead cube, is suspended in an electromagnetic field, floating in isolation from the physical world and presumably shielded from it too. It will be interesting to see how its representation evolves in future films and whether a new quantum filmic imaginary will emerge as it has with other technologies.

Quantum as Imaginary

The lack of quantum-related imagery led us to wonder how a fictional "quantum age" might be visually expressed through the stuff of everyday life. More than styling of course. It's about giving form to new values, beliefs, dreams, hopes, and fears that together form a worldview, or quantum imaginary, a bit like art nouveau, art deco, and streamlining. In *Noise Aesthetics* (2015), designer Lukas Franciszkiewicz explores a visual language for representing fictional products, inspired by ideas from quantum mechanics. Each scene shows an object; *AHA-HUH*, for example, consists of "two mirrored objects, a real one and a simulation, both exist at the same time; The state of the objects is never certain until they are observed." Visually, the objects are represented as though they are located in a liminal reality blurring what is actually there and what might be there. Could a new visual language emerge that captures or even hits at the contradictory nature of quantum mechanics as it plays a more active role in day-to-day life?

Norman Bel Geddes, *Motor Car No. 9* (without tail fin), ca. 1933. Image courtesy of Norman Bel Geddes Theater and Industrial Design Papers, Harry Ransom Center, University of Texas at Austin and the Edith Lutyens and Norman Bel Geddes Foundation, Inc.

Lukas Franciszkiewicz, with Takram, *Device #1: AHA-HUH*, part of *Noise Aesthetics*, 2015.

Lukas Franciszkiewicz, with Takram, *Device #2: Magnetic Noise Feeder*, part of *Noise Aesthetics*, 2015.

Lukas Franciszkiewicz, with Takram, *Device #3: Quantum Fluctuation Bin*, part of *Noise Aesthetics*, 2015.

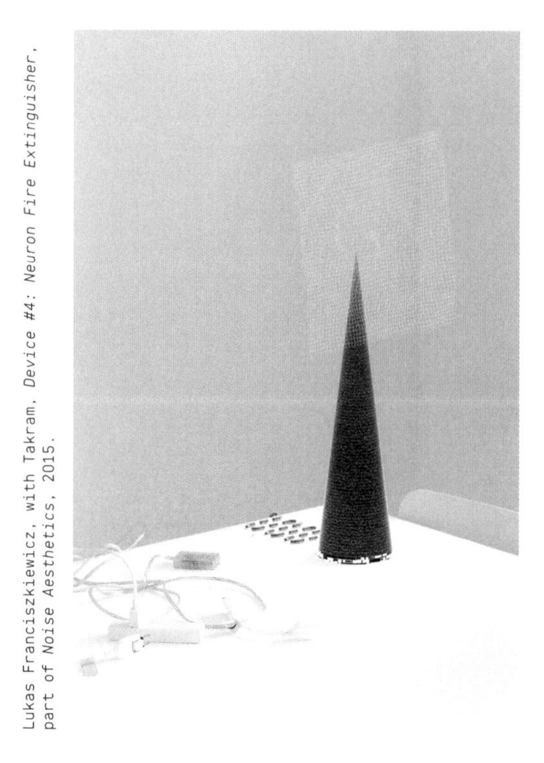

Lukas Franciszkiewicz, with Takram, *Device #4: Neuron Fire Extinguisher*, part of *Noise Aesthetics*, 2015.

Devon Reina, *A Series of Pseudophysical Objects*, *Pencil*, 2019.

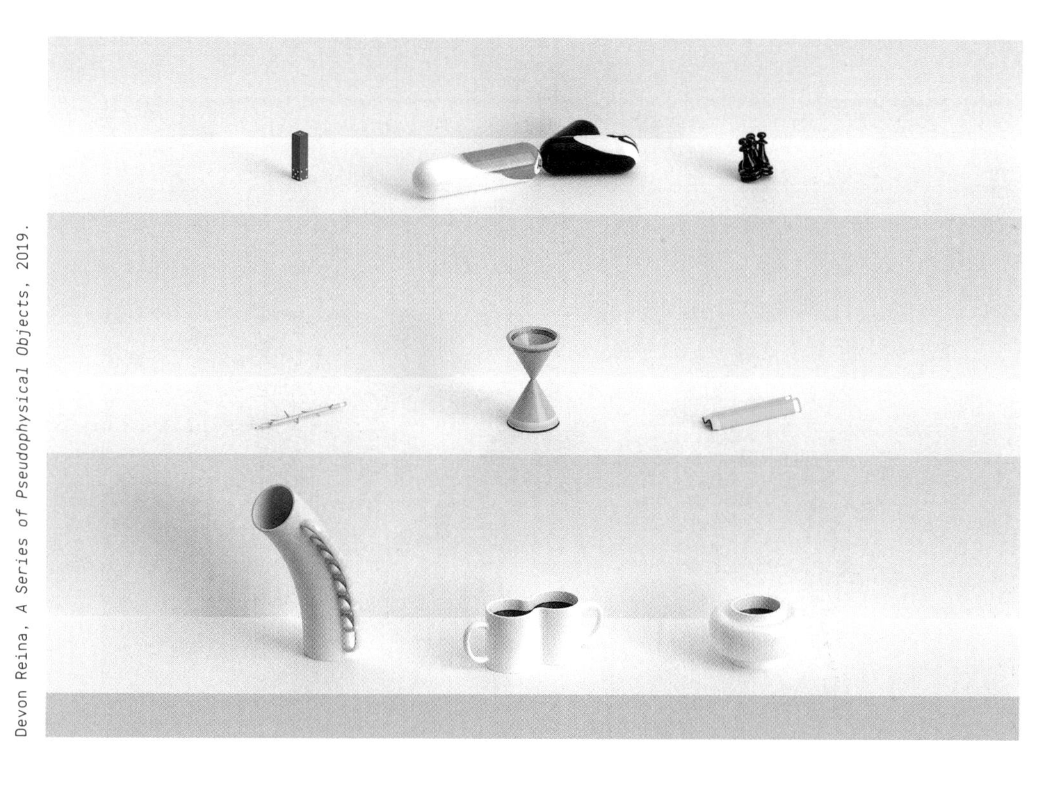

Devon Reina, *A Series of Pseudophysical Objects*, 2019.

There are many projects that although not explicitly linked to quantum mechanics, do begin to suggest how some of its ideas, values, and hopes might find expression in the world of everyday things. *A Series of Pseudophysical Objects* (2019) by Devon Reina, an industrial design student at Parsons in New York City, explored dimensionality through the redesign of everyday objects in one of our classes. A pencil, for example, is rotated about its vertical axis at a slight angle, resulting in a double-cone pencil. Completely nonfunctional, but intriguing. A billiard ball is modeled in motion, striking another, and both are elongated with blurred numbers and distorted markings. We like how it suggests designing from a very different starting point and set of assumptions about objects in space. One can't help but wonder how this might spill over into other areas of life like language, mindset, interactions, gait, posture, greetings, rituals, and so on.

Art studio Troika's sculpture *Dark Matter* (2014) has three different viewpoints showing three different shapes in one object: a square, hexagon, and circle. This and others in a series of perspective sculptures explore tensions between a dualist view on reality. *Squaring the Circle* (2013) consists of a ring suspended in space that transforms from a pure square into a pure circle as the viewer moves around it. *Everything Is and Isn't at the Same Time* (2015), a smaller version of *Squaring the Circle*, continues the theme, but in jesmonite rather than flock.

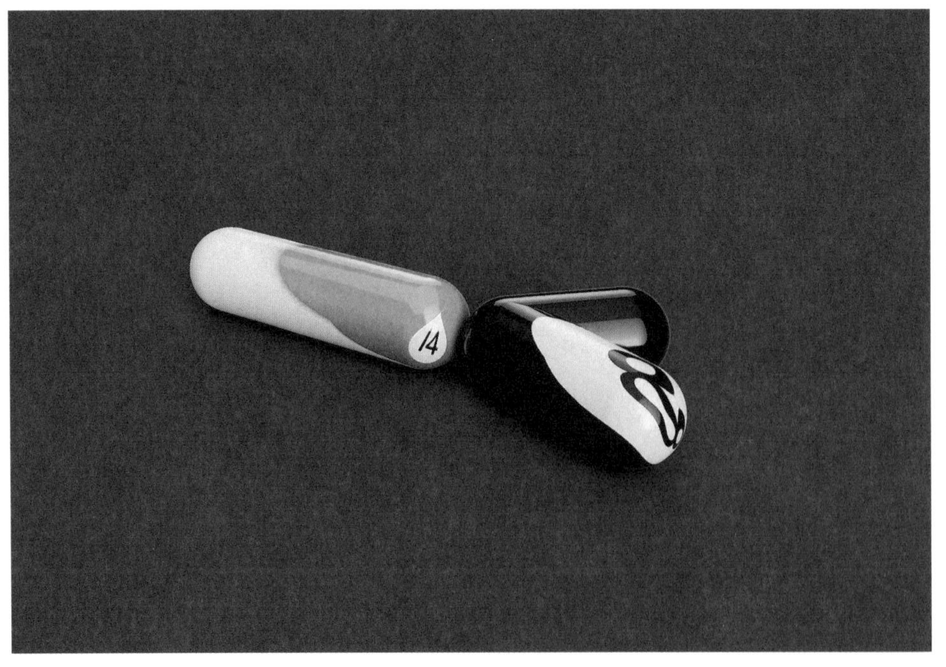

Devon Reina, *A Series of Pseudophysical Objects, Billiard Balls*, 2019.

Troika, *Everything Is and Isn't at the Same Time*, 2015.

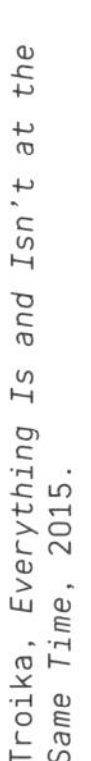

Troika, *Everything Is and Isn't at the Same Time*, 2015.

Troika, *Everything Is and Isn't at the Same Time*, 2015.

In a quantum world, would public buildings take on such forms as we adopted an and-both rather than either-or outlook? A bit like Carolyn Kirschner's *Bim/Sim City*, part of *The London Institute of Pataphysics* (2015-16), which uses a form of three-dimensional, physical anamorphism to represent a model of its headquarters. Like Troika's work, its appearance radically changes depending on the angle it is viewed from.

Underlying these projects is a set of principles related to those in the two-dimensional optical illusions representing impossible three-dimensional figures discussed in the Archive of Impossible Objects. Like these, it is possible to abstract a grammar of three-dimensional anomalies and contradictions that if used as a foundation, could provide the basis for the design language that captures some of the contradictory nature of a speculative quantum material culture. On the surface, this language might begin to resemble cubism. Maybe that's no coincidence, but cubism has a looser, more impressionistic quality. With these projects, although less visually complex, they have a simplicity, precision, and rigor that feels more appropriate for a designed environment. They have a sculptural physicality that renders their effects more compelling, driving home just how much of this stuff happens in the mind.

Although interesting to speculate on what a quantum world or zeitgeist might look like, it is still somehow operating within the "contours of the possible" as defined by a Newtonian mindset. More interesting for us is to start to imagine different belief systems, values, and ideas that transcend a Newtonian way of making sense of the world and use design to give them form.

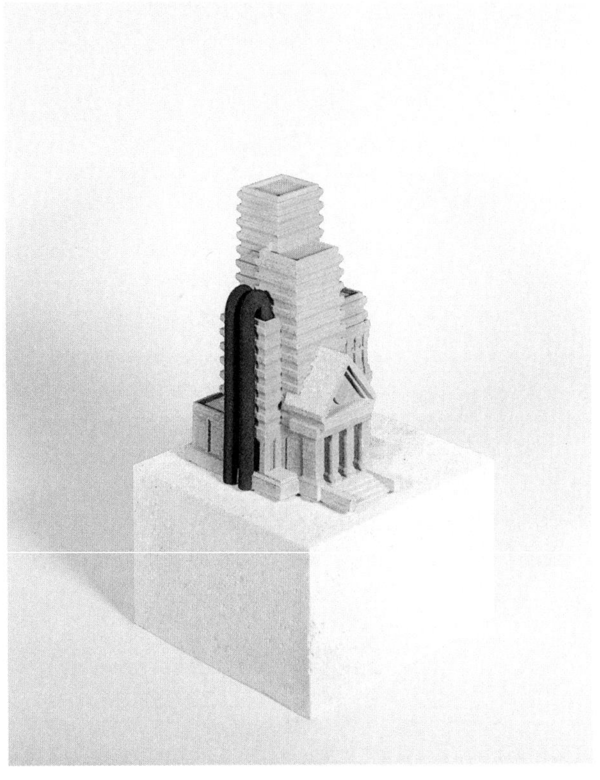

Carolyn Kirschner, *The London Institute of Pataphysics*, 2016. Photo: Matteo Mastrandrea.

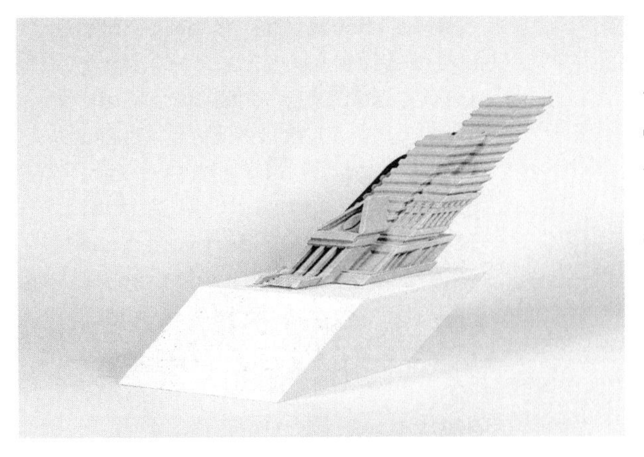

Carolyn Kirschner, *The London Institute of Pataphysics*, 2016. Photo: Matteo Mastrandrea.

Quantum Common Sense

Quantum common sense is an intentional oxymoron linking two worlds that would normally never meet. One concerned with paradoxes that challenge how we make sense of the world and even what the "world" is. The other, a conceptual toolbox for navigating daily life, aligned with how we believe the world works. Current ideas of what constitutes common sense align with what is often called "classical physics," nicely defined by political philosopher Nicolas Harrington in "Time Crystals and the Quantum Mindset":

By Classical worldview, I refer to the "common-sense" assumptions that: (1) everything that happens has a cause; (2) causes precede happenings; (3) "interaction" implies objects are in physical contact; (4) events occur "in time," i.e., events have a beginning, middle, and end; and, (5) time is linear, i.e., there is such a thing as the past, present, and future. In other words, the Classical worldview is the background assumption that the universe operates like some kind of sophisticated machine—like an enormously complicated clock.[21]

Barad's extensive thoughts on Michael Frayn's play *Copenhagen* also hint at what a quantum mindset might be like as well as some challenges that might accompany it: "If we follow the uncertainty principle, we would conclude that we shouldn't presume anything about intentions (since we can't know anything about them) and therefore all we have to base our moral judgements on is our actions. This is what Frayn calls a 'strange new quantum ethics.'"[22] Later, she makes it clear that we cannot make a judgment on observable facts/actions alone but rather can't make judgments at all. This unknowability makes it impossible to judge someone. Barad also points out the risk of this kind of analogical thought—in this case, from physics to psychology—by suggesting we do not need quantum theory to say that certain parts of a person are unknowable. This is the challenge here: how to avoid bolting quantum notions onto existing ideas in ways that are achievable by other means. There needs to be something uniquely quantum, but what exactly is that? For us, it is less about defining "what" quantumness is and more about exploring "how" quantumness might connect with the everyday to lead to new possibilities.

But what does this mean for designers? Especially designers of things, stuff, matter. Ball's *Beyond Weird* offers some intriguing insights by grounding quantum theory in the material world while attempting to show that it is not "weird," just different. For example, using a pen and pencil that have become "entangled" as an analogy to explain what entanglement means in more concrete terms, Ball observes,

If my pen and pencil were entangled quantum objects, it could be that I might examine my pen and find out all that there is to know about it and still not know for sure what color it is—because the color is not entirely there in my pen. Or I could investigate the pen and pencil together, measuring all that there is to know about the two of them as a pair of entangled objects. I can measure, let's say, what "colours" they share between them. And yet if they are entangled then I might end up with complete knowledge—that is, everything knowable—about the pair while being able to say little—or possibly nothing at all—about what each individual element is like, such as what color they are. This is not because I haven't looked closely enough. It is because the entangled pen and pencil may not have local properties. They can't be ascribed individual colours. That is, roughly speaking, what entanglement is like. It is, you might say, a quantum phenomenon through which single objects may be deprived of a definable character.[23]

Hidden Variables: Unknowable Vehicles

The Einstein-Podolsky-Rosen paradox argued that quantum mechanics is incomplete.[24] If proven, this would explain why "spooky action at a distance," or entanglement, is not in fact spooky at all and aligns with known theories. In other words, we are missing something, and quantum mechanics is an "incomplete description of reality." If we could just see it, then it would all make sense within what is already known. Experiments have since proven that there are unlikely to be any hidden variables, and quantum entanglement is indeed spooky, despite defying existing theories. Once entangled, separate objects in two places exist as one—a property known as nonlocality.

But it is the idea of hidden variables in general that appeals to us, and for two reasons. First, there are probably already many hidden variables in the world, and although we think we can see everything and know everything, there are possibly other things at work that we are not even aware of. This calls for a little humility. Second, hidden variables can be used to tentatively align apparently strange phenomena with known explanations, as in Einstein's approach to nonlocality and quantum mechanics. If we could just find the hidden variables, it would all make sense. They serve as a sort of conceptual bracketing, giving something a more ambiguous status; there, but not quite there. To be determined. In both of these senses it is an important idea, and one worth celebrating.

Hidden Variables: Unknowable Vehicles consists of a collection of sixteen vehicle-like objects or maybe one object in sixteen parts. A response to thinking about quantum rather than an illustration of it. Collectively and individually, they don't quite make sense, and that is OK. They are reminders of the presence, possible or otherwise, of hidden variables. A (diplomatic) gift from one group of people to another, from those who have adopted a quantum mindset to those still holding onto a Newtonian outlook, straddling both and comprehendible to both.

We say vehicles, as they have wheels or tracks, and other parts that could contain people or goods, as well as unspecified means of propulsion. But are they vehicles? On closer inspection, many of them don't quite add up. Wheels arranged at ninety degrees to each other. Disproportionately large volumes attached to seemingly inadequate structures and frames. Definitely not streamlined and aerodynamically efficient. Many look as though they would fall over or be impossible to move unless made from currently unknown materials. They are also scaleless. There is something missing; something is not being explained about a world where these belong and "make sense." If vehicles reference the dreams of a particular society, what dreams do these hint at?

On one level, like many of the projects in this book, they are impossible. In this case, reflecting qualities that define aspects of current Western societies, but remain invisible for the most part. Social wrongness is hard to see, even when we know it exists. Yet physical wrongness is almost always obvious to everyone. We have an innate sense of what is possible or not. Using the language of vehicles to make visible the absurdity of our current situation. These vehicles are a gift to a society like ours, where balance is gone, change seems impossible, logic escapes us, and things no longer make sense.

Or are they vehicles for an as of yet unknowable world? Diplomatic gifts intended to serve as reminders of ideas yet to be imagined?

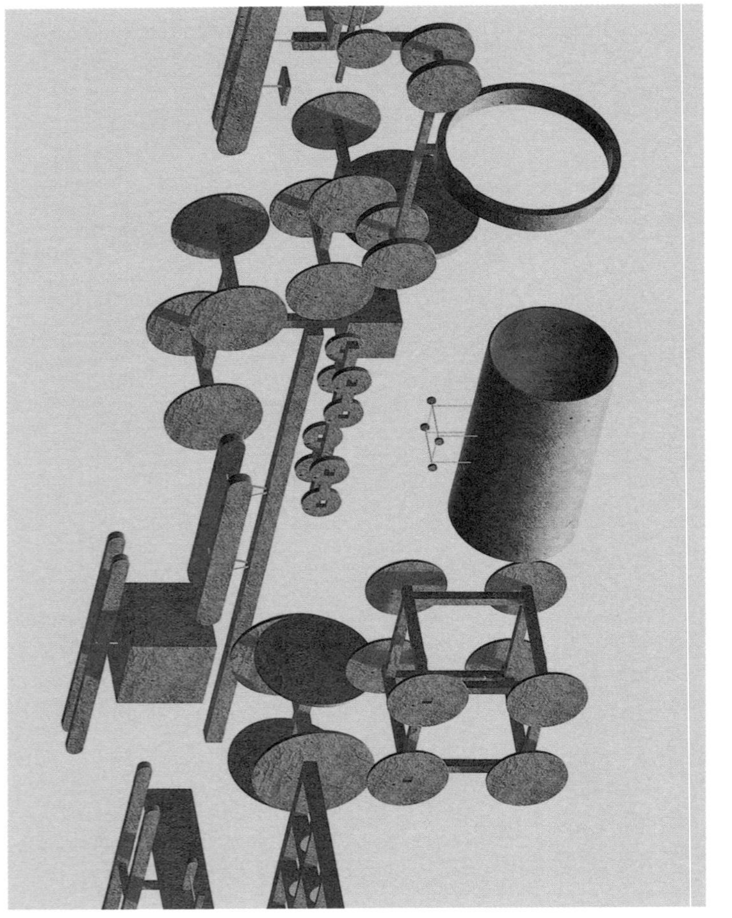

Dunne & Raby, *Hidden Variables: Unknowable Vehicles*, 2023. Detail.

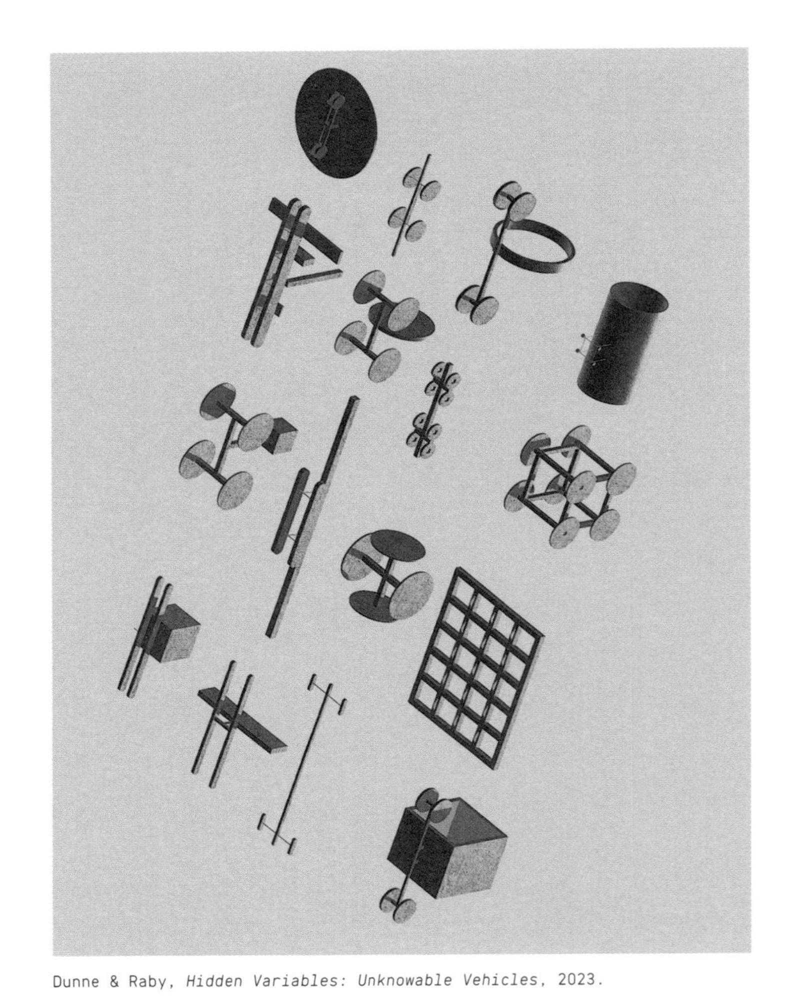

Dunne & Raby, *Hidden Variables: Unknowable Vehicles*, 2023.

Dunne & Raby, *Hidden Variables: Unknowable Vehicles*, 2023. Detail.

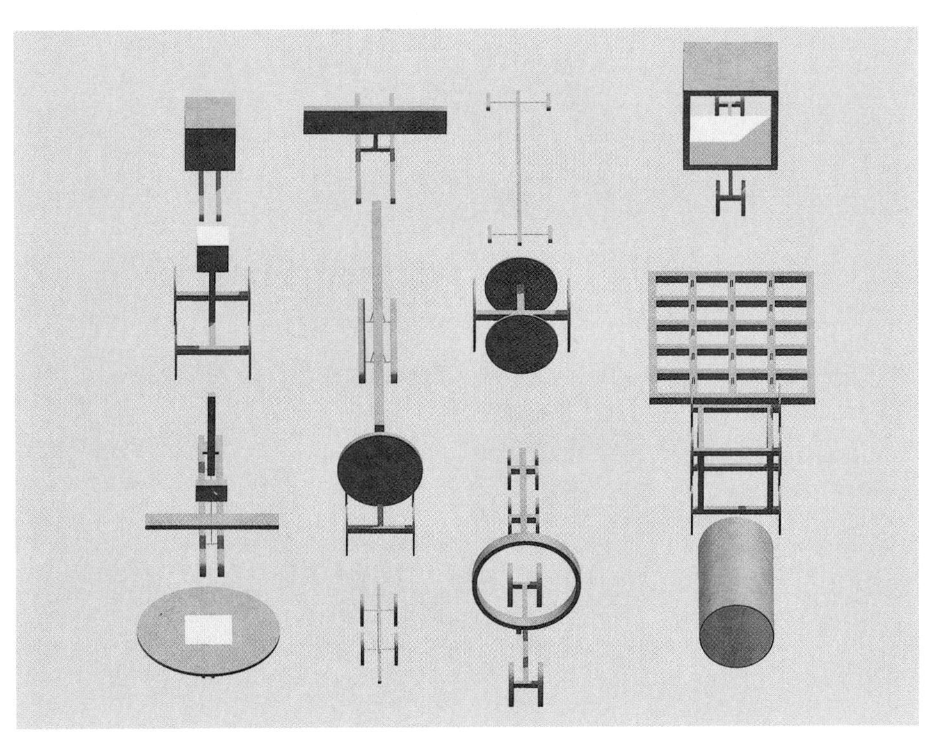

Dunne & Raby, *Hidden Variables: Unknowable Vehicles*, 2023.

4 Unreal

by Design

Chapter 4

Unreal by Design

As the space for imagination has been crushed by the flows of the mediated imaginary, the space of mental autonomy has grown so narrow that we can barely choose what to think about, what to talk about, what to fantasize about.

—Franco "Bifo" Berardi[1]

In our coauthored fictional piece "The School of Constructed Realities," we wrote about our visit to "a new school of design developed specifically to meet the challenges and conditions of the 21st century":

> It offers only one degree, an MA in Constructed Realities. Having sat through the presentations for the open day, we were still a little unclear about its distinctions between real realities, unreal realities, real unrealities, and unreal unrealities, but we were intrigued enough to want to know more.
>
> The school provides a mix of theory, practice, and reflection. There are no disciplines in the conventional sense; instead, students study bundles of subjects. Some that caught our attention were: "Rhetoric, Ethics, and Critical Theory" combined with "Impossible Architecture"; "Scenario Making and Worldbuilding" mixed with "Ideology and Found Realities"; and "CGI and Simulation Techniques" taught alongside "The History of Propaganda, Conspiracy Theories, Hoaxes, and Advertising." Projects are expressed through various forms of reality: mixed, immersive, simulated, unmediated, and so on. Students can also attend the classes "Multiverses and Branding," "The Suspension, Destruction, and Production of Disbelief," "Reality Fabrication: Bottom Up or Top Down?," "The Politics of the Unreal," "Reality: Local Variations," and our favorite, "The Aesthetics of Unreality."
>
> After the presentations we asked the director about the thinking behind the school. They were a little reticent at first, which is understandable knowing the risks associated with relocating design from its cozy home in the old reality-based community to a new one among reality makers, fabricators, and constructors, but they were keen to share. They began by explaining that in their view, for most people today reality isn't working, that it broke sometime near the end of the 20th century: "It's clear that reality only

works for a privileged minority, but designers advocate a realist approach, which means they work within the constraints of reality as it is, for the minority. The school aims to challenge this by making reality a little bit bigger to provide more room for different kinds of dreams and hopes. An important part of this process is giving form to multiple realities, and this is where design comes in."

"We concluded," they said, "that the only way to challenge this unsatisfactory situation was to be unrealistic—to breach realism's heavily policed borders and to fully embrace unreality."

Listening to him, we began to think so too.[2]

Ontological Shock

So how do you "breach realism's heavily policed borders...to fully embrace unreality"?

This is something we constantly grapple with as educators and in our own practice.

If we were to revisit the infamous "futures cone" today, and acknowledge that like many diagrams, it is a simplification with limitations, we might add cones within cones, many of them, all overlapping, nested, different lengths, and different scales—individual, family, work community, nation and so on.[3] Fractal futures. But the one striking difference we would see is the slenderness of the cone(s). It feels as if the various probabilities, potentials, and possibilities it represents have become little more than a sliver. The space we previously felt was unproductive; what lay outside the cone(s) or unreality is now exactly where we need to be looking. Unlike the other sections of the cone, which all, in one way or another, lead back to the present and therefore are versions of it, the external realm offers a genuine space of fluid exploration, a not here, not now.

Our primary tactic for exploring this space is a form of estrangement that attempts to go beyond visual appearances and create a pause that momentarily disrupts the "flow[s] of the mediated imaginary" that Berardi mentions above. It also makes things strange, unfamiliar, and odd, in an effort to slow encounters down and open up new spaces of engagement that gently jolt the viewer out of habitual modes of reading designed objects. Estrangement has many variations: defamiliarization in literature, the A Effect in theater, and cognitive dissonance in science fiction, to mention just a few. It also shares affinities with the weird, strange, absurd, surreal, and so on. Using wrongness in conceptually engaging ways to reveal underlying, overlooked, or invisible assumptions.

To this ever-expanding list, we'd like to add a new kind of estrangement, "ontological shock," which brings further possibilities and depth to the use of estrangement in design. We first came across the term in Kastrup's *Meaning in Absurdity* in a section discussing speculative ontology and the work of Jacques Vallée, who defined it as "the mechanism by which the phenomenon forces an expansion of people's conception of reality towards a worldview where notions previously held to be absurd become intelligible."[4] It's not just a case of providing a fresh perspective on a familiar world but rather grasping fundamentally different logics that allow for a questioning of deeper assumptions that until this moment seemed unquestionable. Instead, the viewer encounters alternative logics, values, and so on, made physical, not as a future proposition or proposal, but as an idea to consider, to toy with, to think about. Its purpose is to expand the imagination, allow for more possibilities, and make reality that little bit larger.

We're not sure that design can ever produce actual, full ontological shock, but it is a potentially helpful way of approaching the idea of estrangement that allows us to go deeper than simply making something appear visually strange. To move beyond materials, function, and use, to question deeper assumptions about what makes sense, of what is possible or not, and why.

As designers, we approach the idea of ontological speculation slightly differently from philosophers. We are more interested in how it might work on the imagination. How it can be used to design useful fictions, and what those fictions might be used for. The idea of design and ontology being entangled in some way is fascinating, and much has been written about it in design studies and related fields.[5] Typically, the label "ontological design" is understood to mean: when we design the world, the world designs us back. This is useful as a frame for critiquing design at a deep level, but less helpful when it comes to exploring new possibilities for design. We are far more curious about what it means as designers to practice at a level that touches on ontology rather than as a lens for thinking about design. And what this might look like, as a form of practice, when design projects start from different ontologies, providing fragments from worlds located in the not here, not now. And how to even begin to put ourselves into radically different relationships to what already exists or might exist.

Too often, critical forms of design practice focus only on the message, often using conventional design languages that drift into parody through a need for legibility by recycling well-used tropes–evil corporation, handmade romanticism, DIY moralism, organically shaped goodness, and jagged punk-like salvage. Here, we are more interested in embodying alternative kinds of sense in the things themselves. Not projecting or conveying, but manifesting in the physical presence of a thing that although physically present, does not belong here or now. As Shaviro puts it, "Fictions, by definition, are works that present us with unreal stories and situations. And yet, these fictions–novels, songs, pictures, theories, and so on–are themselves actual things in the world. They are processes, performances, and objects. They portray unrealities, but they themselves are real."[6] Encounters with objects like this can naturally seem weird, absurd, nonsensical, irrational, and surreal–anything but normal. These are usually negative terms when used in design. But for the kind of work we are discussing here, they are actually necessary.

Ontological Oddities

> If a physician is someone who *practices* medicine, perhaps a metaphysician ought be someone who practices ontology. Just as one would likely not trust a doctor who had only read and written journal articles about medicine to explain the particular curiosities of one's body, so one ought not trust a metaphysician who had only read and written books about the nature of the universe.
>
> –Ian Bogost[7]

The first object in our Archive of Impossible Objects (chapter 2), a spear from O'Brien's *The Third Policeman*, has a point so absurdly sharp, that it is felt six inches before it can be seen by the naked eye. Its existence (in the story), rather than any particular message, is what is important here. And of course the world it belongs to: one similar to ours, but a little different, a little more absurd and irrational. In previous books when we have encountered objects like this, made from words, that can only exist in text or image, we have quickly backed away to explore what can be done in the world of actual objects. But in the context of working with unreality, it might at last be interesting to explore how the qualities that make some of these textual and imagistic examples so interesting can also be examined in the world of things, as objects. To develop design analogues that embody contradictions, paradoxes, and impossibilities, and sit at the edge of what makes sense.

On encountering objects like these, what we think of as ontological oddities, the question that needs to be asked is a simple one: Is it interesting? And by that we mean, does it expand the imagination, challenge the intellect, expose hidden assumptions, or offer new insights? Basically, does it open new possibilities for thought and imagining? If not, then it fails. In some ways objects like these only have one job to do, but they need to do it extremely well, and that is where the craft lies. This is not an easy task, and more effort is needed to understand how to do this well. Too often, the message is valued over materialization and affect. Or it stops with materiality—renewable, sustainable, novel versions of what is already known rather than new kinds of knowing.

In this role, the designer's task is to give form to fragments from a multiverse of possible worlds that contribute to a culture of imaginative alterity materialized in ways that engage the mind by challenging it, shifting its focus, arresting it, motivating and inspiring. Raising awareness that if reality is not given but instead made, then it can be unmade and remade. This is not just about reimagining everyday life—there are plenty of examples of this; it is about using unreality to question the authority of a specific reality in order to foreground its assumptions.

It is hard to find design examples that do this, and we frequently turn to fine art, where there is more freedom and fewer expectations that things should "make sense" within conventional frameworks. The qualities we find interesting in many of these works are side effects, and possibly unintentional ones, where in nearly all cases, the creator's intentions were probably different. Nevertheless, they still serve as valuable examples of what objects embodying fictional ontologies might look like.

But let's start with design. In the project *For the Rest of Us* (2018) by Hank Beyer and Alex Sizemore, there are a number of seemingly nonsensical objects: a keyboard made from ice, and various computers made from sandstone, coal, honey, peat, and lard. Clearly they are not functional. They belong to a "possible world" where seemingly familiar objects exist in different ways to those in ours or familiar objects are used to suggest other logics that current design language does not allow for. The designers' interest was in drawing attention to working with local materials and moving away from more harmful materials such as plastic, not just on a functional level, but an emotional one. As they say in an interview about the project, designers "have a responsibility to explore these seemingly impossible realities in order to push our world towards more meaningful, physical futures."[8] They are not trying to show us what

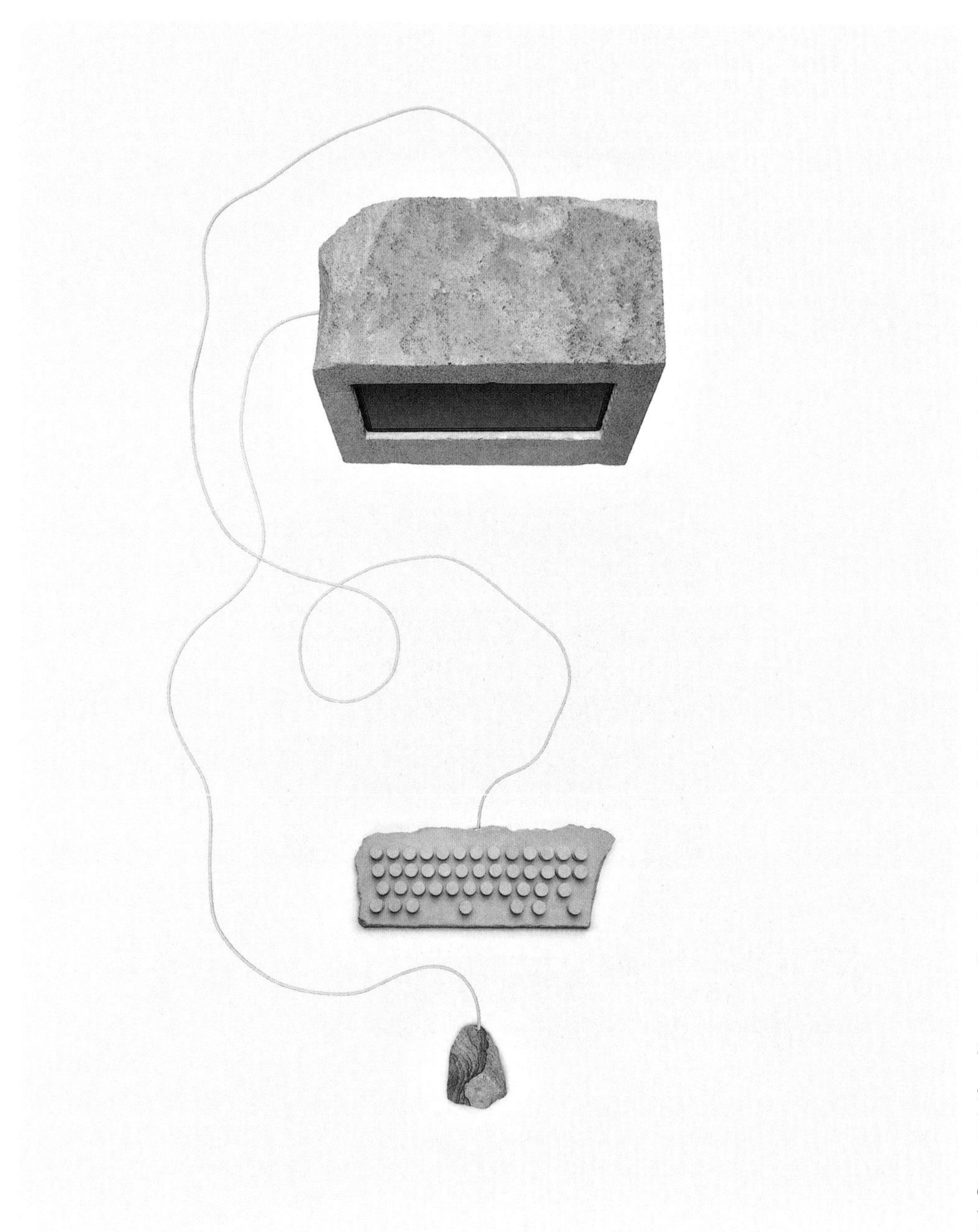

Hank Beyer and Alex Sizemore, *Computer Made from Sandstone*, 2018, part of the series *For the Rest of Us*.

Hank Beyer and Alex Sizemore, *Computer Made from Coal*, 2018, part of the series *For the Rest of Us*.

our world would look like if we embraced these values but instead hint at worlds that open the viewer's mind to different possibilities and values, concerned with how we use materials today. They are more about undoing what we think objects are, breaking with current mindsets, in preparation for whatever is to come next.[9]

[FEx631.7]-[9oi], part of the *[KAJOLA]* (2021) collection by Yussef Agbo-Ola, works in a similar way. The project consists of series of trainers assembled from biomaterials including agricultural waste, natural fibers, and root extracts, such as red jasper dust, desert silica, algae fiber, violet volcanic ash, peat, coconut shells, almond milk pulp, pyracantha berries, turmeric, Ce[U]12i, and cotton fiber, to create living footwear using plant skins that provide medicinal properties as well as embracing decay. Rather than attempting to show how they would be made or used in this reality, they hint at an imaginative leap that will need to be made if biomaterials are to be truly embraced—messy, roughly assembled, imprecise, decaying, and probably smelly too. Embracing more sustainable materials is not just a practical choice but will also require a significant cultural and aesthetic shift. These could easily be dismissed as aesthetic studies, yet we believe they serve other purposes, embodying a very different understanding of what a new, less harmful material world might look like.

Yussef Agbo-Ola, Olaniyi Studio, *[FEx928.1]-[9oi]*, part of the *[KAJOLA]* collection, 2021. Photo: Yussef Agbo-Ola.

Yussef Agbo-Ola, Olaniyi Studio, *[FEx645.9]-[9oi]*, part of the *[KAJOLA]* collection, 2021. Photo: Yussef Agbo-Ola.

Yussef Agbo-Ola, Olaniyi Studio, *[FEx747.0]-[9oi]*, part of the *[KAJOLA]* collection, 2021. Photo: Yussef Agbo-Ola.

Olivia Bax, *Suck*, 2020. Polystyrene, funnel, handle, paper, ultraviolet-resistant polyvinyl acetate, household paint, and plaster. 39 × 29 × 35 cm. Courtesy of the artist. Photo: Tim Bowditch.

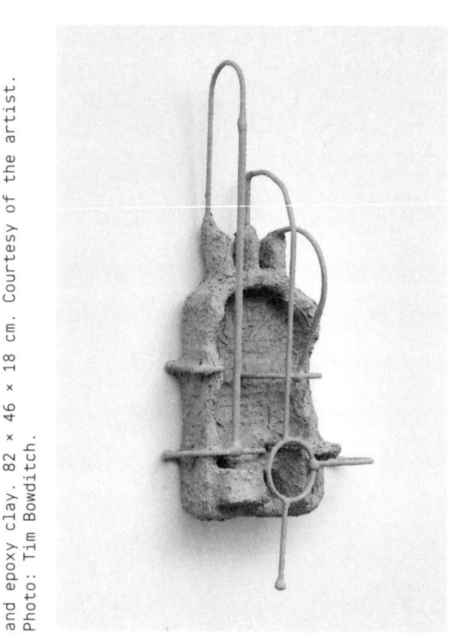

Olivia Bax, *Home Range*, 2021. Steel, polystyrene, newspaper, ultraviolet-resistant polyvinyl acetate, household paint, and epoxy clay. 82 × 46 × 18 cm. Courtesy of the artist. Photo: Tim Bowditch.

Sculptor Olivia Bax makes functional-looking objects from steel, polystyrene, cardboard, epoxy clay, newspaper, plaster, polyvinyl acetate, household paint, and varnish that look like components from slightly strange engineering projects or specialist furniture, yet they do not meet any practical needs. They are wobbly, irregular assemblages that resonate with modernist structures and maintain links to the functionalist mindset underpinning so much of our built environment. At first glance, they may appear similar to a recent genre of design art where chairs, tables, shelving units, lights, and other familiar objects are reimagined as blobby, organic, freestyle forms. But whereas these remain within the constraints of existing object typologies and behaviors, Bax's objects defy easy categorization and open up more possibilities for thought. Or again, maybe for undoing our current thinking in relation to everyday objects.

Bodys Isek Kingelez's models, or "extreme maquettes" as he called them, work very differently. At first glance, they suggest functional buildings and settlements, almost familiar, if a little fanciful. But on closer inspection, they don't quite add up; they are assemblages of packing materials along with other bits and pieces. Again, like so many other projects here, they are not intended to show us what his imagined world would actually look like but instead to evoke a spirit of joy and celebration that could only arise in alternative systems to those shaping our current realities. They are ultimately about celebration. The titles of his pieces are important too, and suggest new civic functions, types of social housing, and institutions for existing and new towns. Their purpose seems to be about sustaining the imagination, to keep it alive and nourish it, so that it remains open to other possibilities. These objects establish a dialogue between the real, fictional, possible, impossible, unreal, and so on.

Bodys Isek Kingelez, *Mongolique Soviétque* [*sic*], 1989. Cardboard and foam, 62 × 63 × 40 cm. © Bodys Isek Kingelez. Courtesy of the Jean Pigozzi African Art Collection. Photo: Maurice Aeschimann.

Bodys Isek Kingelez, *Zaïre*, 1989. Paper, cardboard, foam board, and plastic, 76 × 83 × 97 cm. © Bodys Isek Kingelez. Courtesy of the Jean Pigozzi African Art Collection. Photo: Maurice Aeschimann.

There are many fascinating examples to be found in architecture, usually explored through drawings, model making, and more recently, computer-generated images. One group that consistently engages with the ontological aspects of design—not just at the scale of buildings, but objects—is ADS4 at the Royal College of Art in London. A teaching unit that uses architecture as a medium for engaging with "a world where reality itself appears to have become unreal." Unit themes include Null Island, Postproduction: Manual for Redesigning Reality, Super Models, and in 2021, Legal Fictions, which explored questions around the "ethics of the possible" in a situation where "our world is best understood as a cartoon—an irrational landscape of plausible impossibility (as Walt Disney put it) where anything can happen, yet where certain things reliably reoccur."[10]

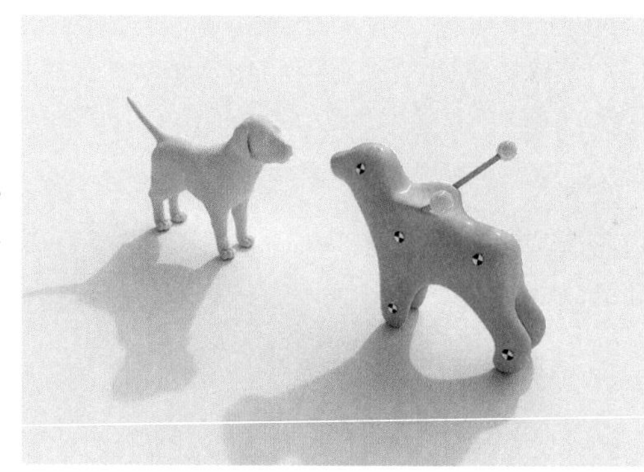

Dominic Oliver, *Owner Introduces Their Dog to a Proxy Future Pet*, part of the series *Proxy Objects*, 2021.

Dominic Oliver, *Bathing with Possible Sea Creatures*, part of the series *Proxy Objects*, 2021.

Proxy Objects (2021) by Dominic Oliver starts with the idea that "given enough time and space, anything that can exist, will exist." The project "questions whether claiming to know is ethical. By anticipating the eventual existence of all possible entities, the project creates a scenario in which Occam's Razor, the apogee of reductionism is futile. The proxy—an object that 'stands in' for its yet to exist referent, develops a language of not knowing, intentional in its inaccurate representation, communicating its lack of communication."[11]

Returning to design, the motivations today for presenting alternative worlds are often related to sustainability. Of showing a glimpse into a version of our world where the human-made elements are less harmful, or more sympathetic to other forms of life and the planet itself. But many of these are still about sustaining existing ways of being, just less harmfully. From biomimicry to biodesign to more-than-human design, the goal is to carry on, but more sympathetically. The focus is on reimagining materiality rather than ways of being in the world.

Much work is being done, for example, on how mycelium-based materials could replace plastics and polystyrene foams for packaging.[12] While these are undoubtedly an improvement on the current situation where packaging generates vast amounts of waste, toxic microparticles, and landfills, in some ways these ideas also sustain existing behaviors, although they allow us to do it less harmfully. Perhaps this is necessary if these ideas are to happen within existing reality, in the "here and now."

A more radical approach can be found in *Entangled Life: How Fungi Make Our Worlds, Change Our Minds, and Shape Our Futures* (2021) by Merlin Sheldrake. He suggests adopting a nonhuman mindset based on a fungi worldview. This of course is not meant to be taken literally but more as a thought experiment to see what new insights might be gained by entertaining mycelial ways of being in the world. There are many projects that look to the world of mushrooms for answers to our current environmental problems and challenges. For example, the Infinity Burial Suit made by Coo and designed By Jae Rhim Lee, made from mushroom spores that help decompose the body and filter toxins so that the surrounding soil is unharmed.[13] And Anna Tsing's *The Mushroom at the End of the World* explores what can be learned from a rare mushroom, matsutake, about living differently on this planet. But Sheldrake is suggesting something more radical: "How different would our societies and institutions look if we thought of fungi, rather than animals or plants, as 'typical' life-forms?"[14] Using design to materialize how this might look in terms of everydayness could be an interesting way to prompt further thought. A sort of stepping stone into the unknown.

The Department of Seaweed, a design and research studio based at Aalto University in Helsinki, is an example of how design can materialize alternative values without prescribing how we should live. Opening up conversations and inspiring interest in what alternative ways of being in the world might look like. Rather than designing everyday objects such as furniture or clothing made from seaweed-based materials, the studio creates abstract objects with names like *Corpus Maris II*, *Hidaka Ohmu*, and *Kombu Nudibranch*. The motivation behind these is to sidestep conversations about functionality, economics, and a fit with how the world is now, and instead create wonder, opening the mind to a world that is not just made from different materials but also from new values and belief systems. They don't show or even tell ,but rather evoke, and that is their value.[15]

But there are always limits to what can be imagined. In Charles L. Harness's short story "The New Reality," ontologists are charged with ensuring reality remains stable by suppressing any research that might

Julia Lohmann in the Department of Seaweed, Victoria and Albert Museum, 2013. Photo: Petr Krejci.

Julia Lohmann, *Oki Naganode*, Victoria and Albert Museum, 2013. Photo: Petr Krejci.

lead to devices that somehow transcend the human umwelt to reveal the "thing in itself" or the "really real," which would result in madness as the human brain struggles to deal with reality in its raw state.[16] How this would happen we are not sure. One might think surrealism would provide many examples of this, but they are more often manifestations of inner fantasy worlds. Their strangeness is the end goal rather than a means to an end, a way to draw people into a more contemplative relationship to the world through the object.

Filippo Nassetti, *Spaceship Generated by Midjourney in the Style of Piero Della Francesca*, 2022.

Ron Luther, *Emergent_Interface #79*, made using Midjourney, 2022.

If we are talking about going beyond the human imagination, then one would think that machine learning programs like DALL-E and Midjourney would be an ideal source of images hinting at strange worlds and impossible objects. But there are limitations here too. They depend on preexisting images, usually from the internet, so there is always something a little familiar about them; in many cases, it has become something of a style. And even with the most advanced programs, they depend on word prompts, which again, limit what is produced to the words used. Yet they can still surprise. Many of these images focus on scenes, such as those of Filippo Nassetti, who in one set of experiments, asked Midjourney to visualize spaceships in the style of Hieronymus Bosch, Canaletto, Giovanni Battista Piranesi, William Turner, Francisco Goya, and Piero della Francesca.

When software like this is used to generate objects, it is frequently even less successful as its source material is often obvious, especially when grafted onto known and easily recognized object typologies. But one set of experiments stands out. An Instagram page called Emergent_Interface consisting of examples of strange interfaces that retain enough of the object typology to work, while escaping expected material logics. In this case, they often suggest stonelike technologies hinting

Eric Tabuchi, untitled, part of the series *Third Atlas*, 2023

at abstract interfaces. *The Third Atlas*, a project by photographer Eric Tabuchi, uses Midjourney to generate mainly architectural structures, but occasionally objects too. Many of the images manage to break with an already familiar Midjourney aesthetic and suggest photographs of scenes that never happened—a number of which involve gatherings of people studying unknown but intriguing models. Beyond their visual strangeness, the models suggest other kinds of knowledge. Rather than replacing the human imagination, machine-generated imagery like this might raise the bar for designerly imaginings.

Treading Lightly in a World of Worlds

There is much talk in design about the "more than human," "decentering the human," "multispecies world-building," and other related ideas.[17] But what does this actually mean for design practice? All too often it appears to simply be a welcoming of nonhumans into a still-human world, an expanding of the range of stakeholders in a project to include different species, or even the extension of codesign strategies to the other than human. This is all good, of course, but it still feels like a one-world world, a human world.

Having spent many years living and working in cities when we moved to the United States, we became fascinated by its many large areas of land rarely entered by humans. They feel like separate worlds where a different, nonhuman logic prevails. To spend any time in these places requires a great deal of preparation. You begin to see yourself not as human but rather as stuff, material, part of the environment. At the same time, we moved from the city to a woodland where the walls of the house define a threshold between worlds. Inside is the human part, but directly outside is a multispecies zone, shared with many other creatures—bears, beavers, eagles, bugs, mycelia, and the most extraordinary fungi.

Then, in spring 2022, we taught a class called Who Comes after the Human? with cultural theorist Dominic Pettman for a mix of design, art, and liberal studies students. It explored how design and theories concerned with the more than human might enter into dialogue and allowed us to think more abstractly about our experiences in the wilds. One idea stayed with us: that we can only ever experience an edited version of whatever is out there—what we think of as reality—filtered through our senses, and that each life-form experiences its own unique world based on its senses and cognitive apparatus. The main point being that as humans, we assume there is only one world, and that all animals share that world. As Haldane asks in *Possible Worlds and Other Essays*, "How does the world appear to a being with different senses and instincts from our own; and if such beings postulated a reality behind these appearances, what would they regard as real?"[18]

We began to wonder what it would mean if we took von Uexküll's idea of an unwelt more seriously and built on the ideas behind the "Object 10: A Human Imagined through a Generalized Nonhuman Unwelt" section from the Archive of Impossible Objects in chapter 2. Rather than thinking about how to invite nonhumans into a human world, maybe first we need to undo how we imagine the human, at least in terms of how we appear or are present in nonhuman worlds—how we are transformed, materially and conceptually, if we acknowledge that nonhuman worlds exist, based on different senses, whether olfactory, electric, seismic, magnetic, or auditory. What would it mean to enter as designers into the highly speculative space of nonhuman ontology? Where things that are invisible to us, for example, might beconcrete and tangible in other worlds, and what is seemingly solid to another animal might be imperceptible to us. In some situations, the wavesand air disturbances we create as we move through space might be as important as our meatiness, or more important. From a modernist point of view, we think of air as empty space, but it is of course fulsome, teeming with life and different materials at a microscopic level.

In a world of many worlds, a teapot might be less present than the air around it. From a nonhuman perspective, objects that we give distinct identities to through language—like teapot, steam, or air—might become unified in ways that fuse words into new object identities.

Electric fields, sweat, warmth, and other materialities invisible to humans might modify the form of the body for those who experience the world through different senses, so the human appears more like an atmosphere than a solid mass. We read that bears can smell food up to a mile away. Does this mean that we extend through their world as delicate molecular strands entering their bodies to become entangled with them? Instinctively, it feels like these traces should be represented in a delicate way—floaty, barely there. But that is a human perspective. Although it was counterintuitive, for our project *Designs for a World of Many Worlds: After the Festival* for the NGV Triennial 2023, we began to explore materials that would suggest solidity.

Of course, attempting to materialize how humans are present in nonhuman sensory worlds is full of contradictions. It is impossible to truly imagine what this would be like; we are always limited by our human imagination and senses.

But this project is not about seeing the world as a bat would. Nor is it about trying in a shamanistic way to become other than human. It is about shifting perspective and seeing ourselves and the human world differently. And hopefully, making us better cohabitees and neighbors by helping us appreciate the nature of different nonhuman worlds as well as how we might be present in them.

In order to move beyond simply illustrating this idea and ground it in the everyday, we framed the project as a festival—one that celebrates the contradictions of working with nonhuman umwelten while challenging what we think a festival is or could be: muted, quiet, slow, careful, and deliberate. There were many things we could design for a festival—costumes, movement, and sounds—but one aspect that caught our attention was the many kinds of poles that people carry in festivals, such as staffs with effigies on them. Often made locally, they can be highly crafted and made to last, DIY, or made from more transient biomaterials. In between festivals, the poles and other accessories, like headwear, footwear, and backpacks, might live in the home, passed down through generations, serving as daily reminders—amid tables, chairs, and other domestic objects—that the human world is just one of many. Like the best utopias, the festival is not meant to be realized; it is intentionally impossible, and that is its value. The fact it is not real allows us to move beyond practicalities to discuss other ways of seeing the world, made tangible through design.

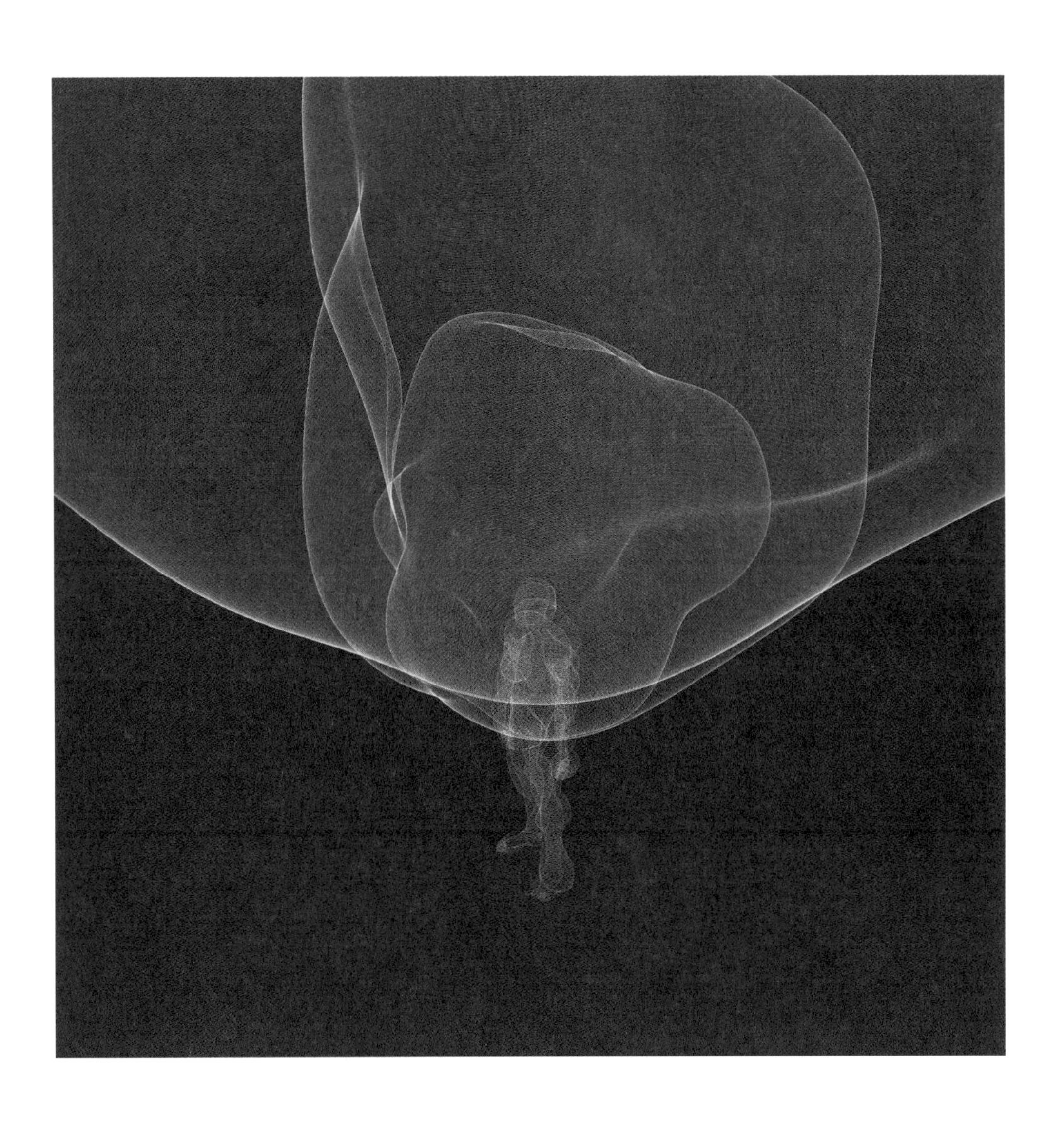

Dunne & Raby, *UW_01*, part of the series *Designs for a World of Many Worlds: After the Festival*, 2022-23. Illustration: Franco Chen.

Chapter 4: Unreal by Design

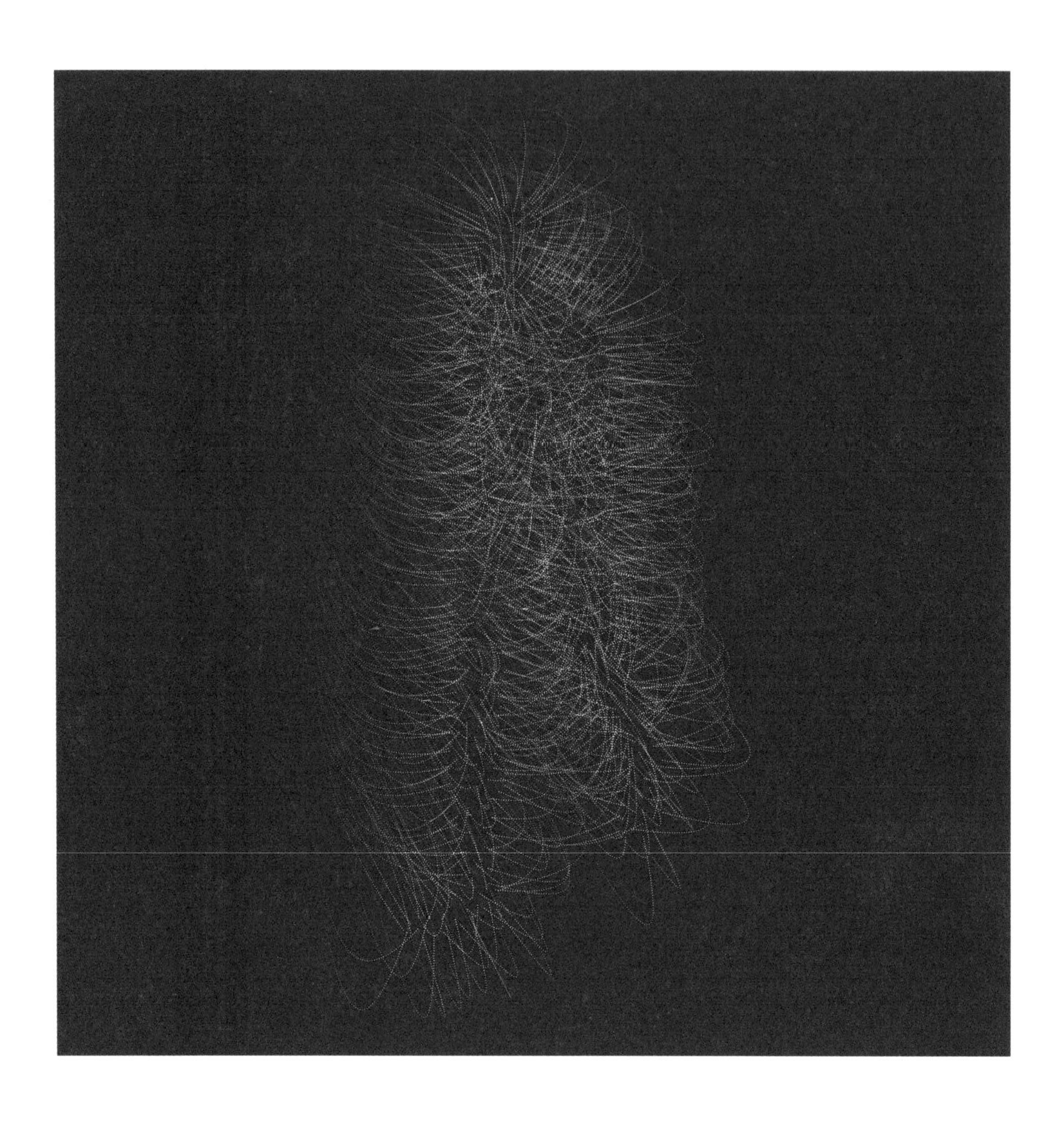

Dunne & Raby, *UW_02*, part of the series *Designs for a World of Many Worlds: After the Festival*, 2022-23. Illustration: Franco Chen.
Dunne & Raby, *UW_03*, part of the series *Designs for a World of Many Worlds: After the Festival*, 2022-23. Illustration: Franco Chen.
Dunne & Raby, *Plume_01*, part of the series Designs for a *World of Many Worlds: After the Festival*, 2022-23. NGV Triennial 2023 at NGV International, Melbourne. Photo: Kate Shanasy.

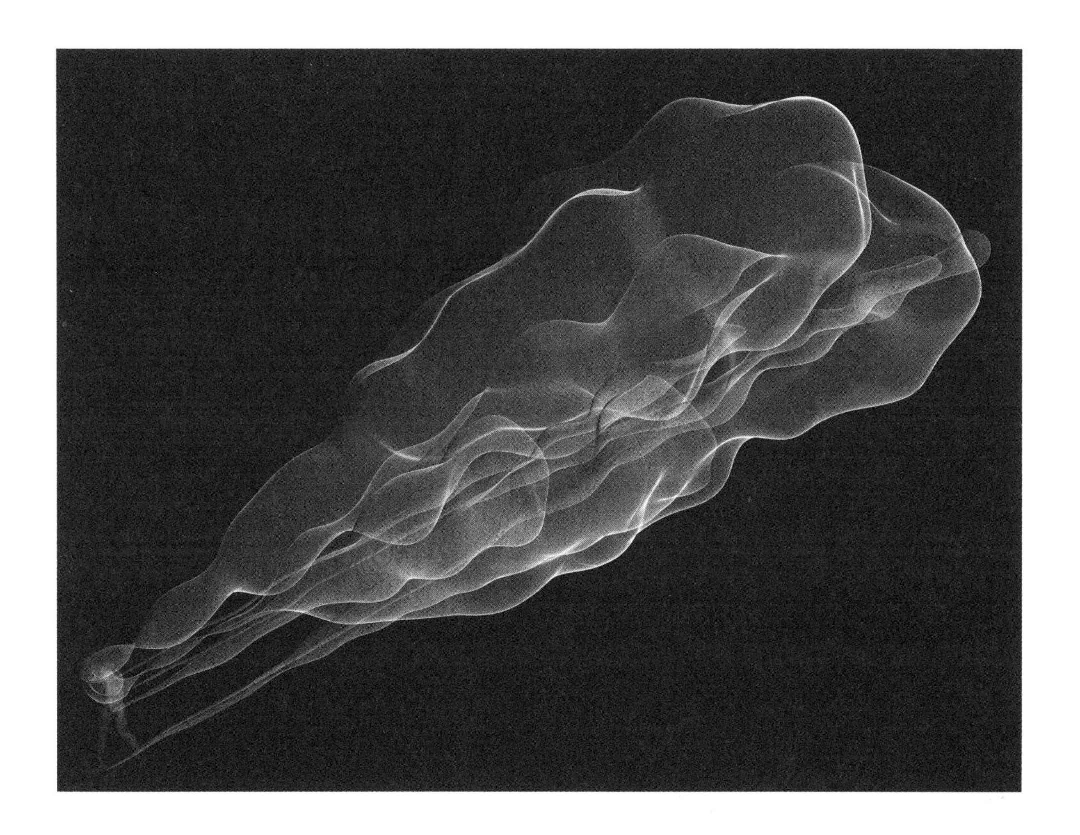

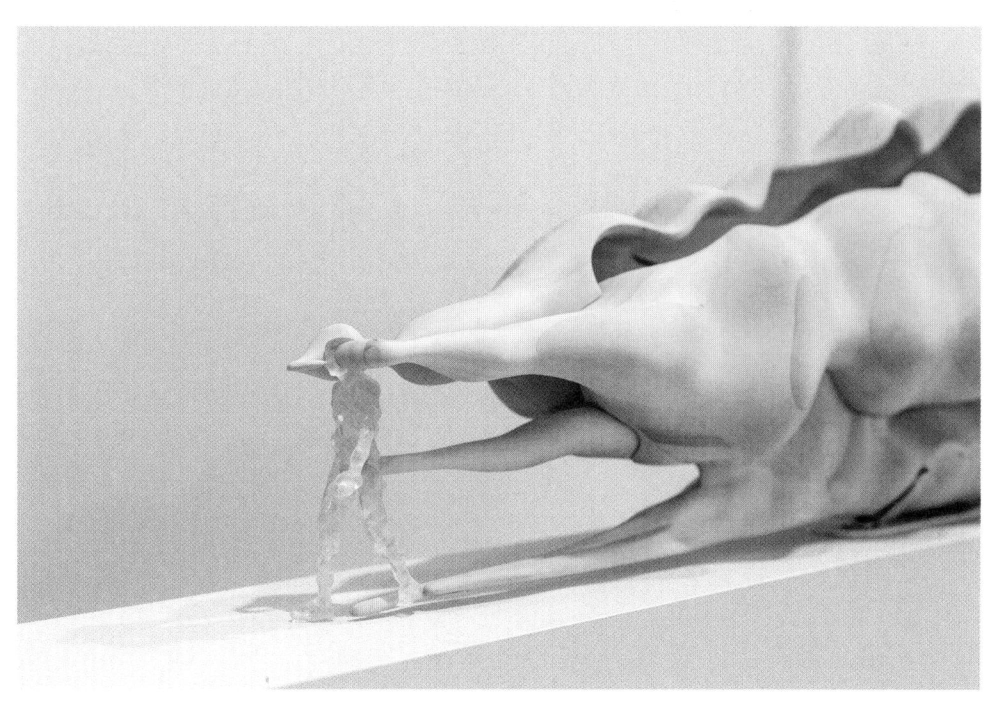

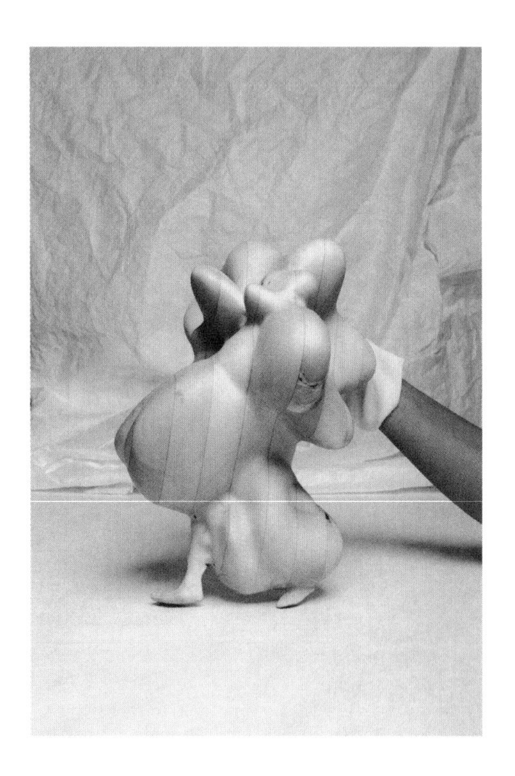

Dunne & Raby, *Plume_01 (detail)*, part of the series Designs for a *World of Many Worlds: After the Festival*, 2022-23. Photo: Kevin O' Connor.
Dunne & Raby, *Umwelt Figure*, part of the series *Designs for a World of Many Worlds: After the Festival*, 2022-23. Photo: Alain Pottier.
Dunne & Raby, *Teapot-Air*, part of the series *Designs for a World of Many Worlds: After the Festival*, 2022-23.
Work in Progress, International Programme for Visual and Applied Arts Open Studio, 2022. Photo: Jean-Baptiste Béranger.

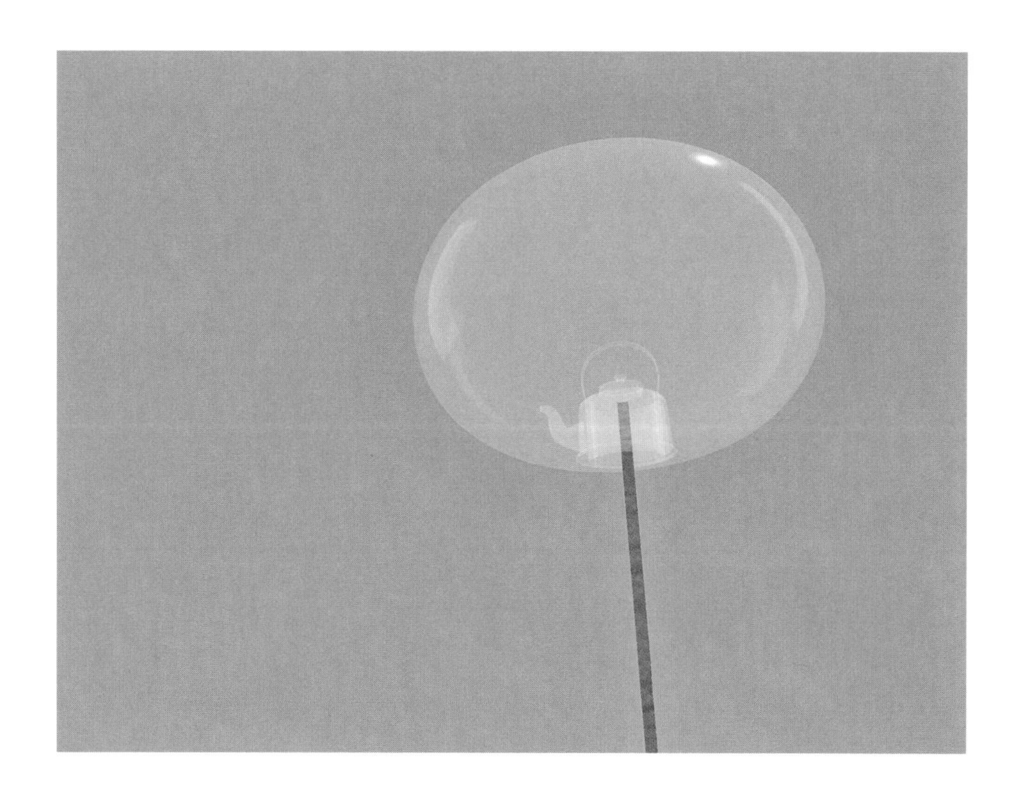

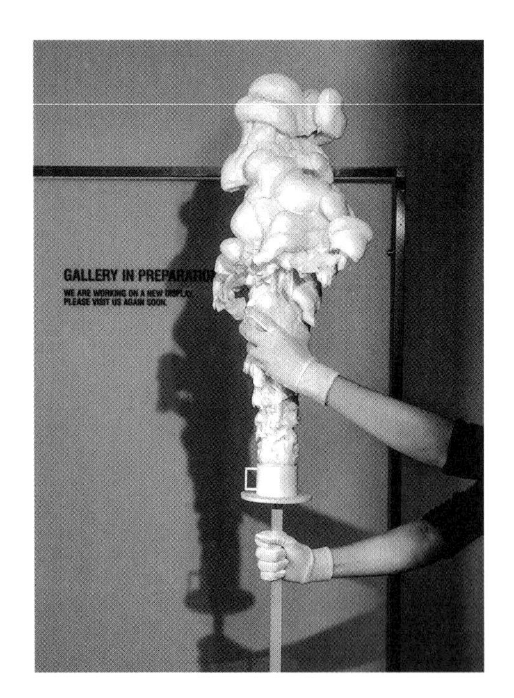

GALLERY IN PREPARATION
WE ARE WORKING ON A NEW DISPLAY.
PLEASE VISIT US AGAIN SOON.

Dunne & Raby, *Cup-Heat*, part of the series *Designs for a World of Many Worlds: After the Festival*, 2022-23.
Dunne & Raby, *Plume_01*, part of the series *Designs for a World of Many Worlds: After the Festival*, 2022-23.
Dunne & Raby, *Cup-Heat*, part of the series *Designs for a World of Many Worlds: After the Festival*, 2022-23. Photo: Alain Pottier.
Dunne & Raby, *Hat*, part of the series *Designs for a World of Many Worlds: After the Festival*, 2022-23.
Dunne & Raby, *Cloud*, part of the series *Designs for a World of Many Worlds: After the Festival*, 2022-23.

So what does it mean to design for a world of many worlds?

For a start, all worlds are equally real. Our senses edit what is out there, as do those of every creature.

Some life-forms live in worlds where the human body shape is meaningless. We appear more like clouds, atmospheres, or energy fields as our meatiness fades into insignificance.

Our breath forms chemical swirls drifting through multiple umwelten. Strands of you stretch for miles, caressing the nervous systems of innumerable life-forms.

Every footstep sends minute waves through the ground, disturbing life within.

A click of the fingers travels further than one might think.

If traveling with others, you become a many-legged, many-armed human mass.

Our communication media forms dense, electric features in landscapes navigated by birds and other migrating creatures.

The sound of a boat creates underwater acoustic masses as solid to some life-forms as mountains are to us.

And ultimately, once we leave the human world behind, we become little more than biomass to be recycled and processed by a multitude of life-forms for which we are simply calories, nutrition, raw material, or a home.

Dunne & Raby, *Plume_02*, part of the series *Designs for a World of Many Worlds: After the Festival*, 2022-23.
Dunne & Raby, *Plume_02*, part of the series *Designs for a World of Many Worlds: After the Festival*, 2022-23. NGV Triennial 2023 at NGV International, Melbourne. Photo: Kate Shanasy.

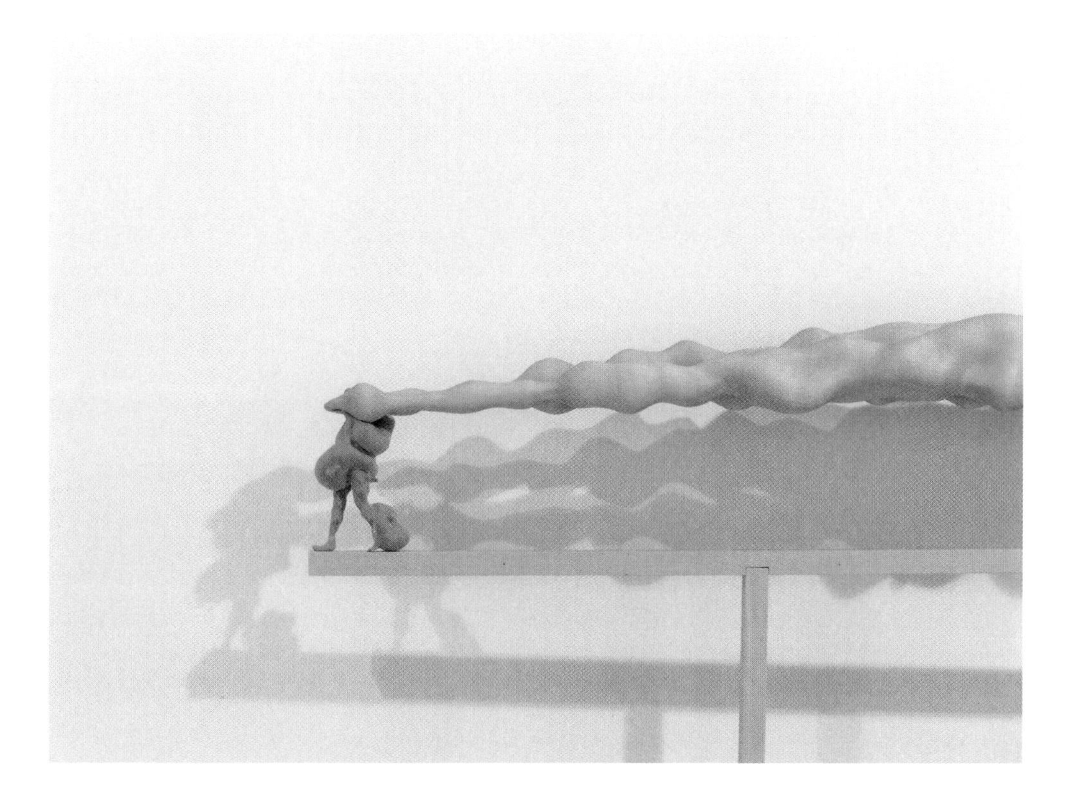

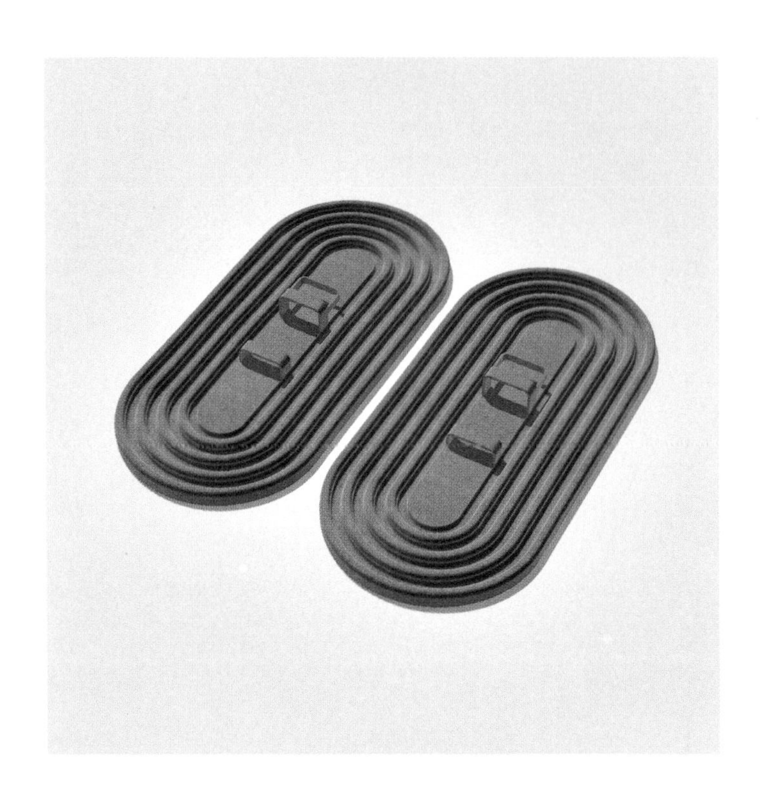

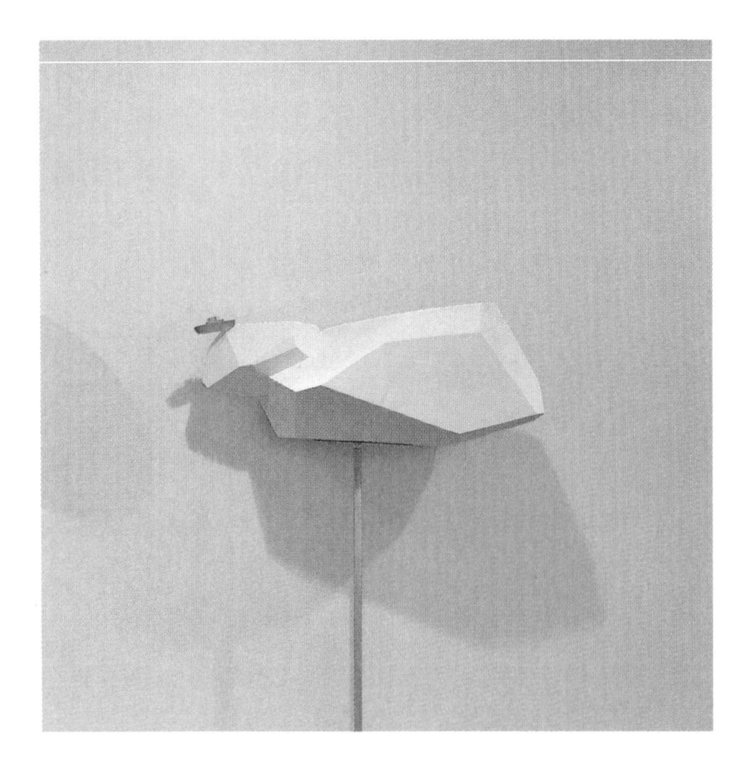

Dunne & Raby, *Felt Shoes*, part of the series *Designs for a World of Many Worlds: After the Festival*, 2022-23.
Dunne & Raby, *Boat Sound Mass*, part of the series *Designs for a World of Many Worlds: After the Festival*, 2022-23. NGV Triennial 2023 at NGV International, Melbourne. Photo: Sean Fennessy.
Dunne & Raby, *Click*, part of the series *Designs for a World of Many Worlds: After the Festival*, 2022-23. NGV Triennial 2023 at NGV International, Melbourne. Photo: Kate Shanasy.

Dunne & Raby, *Human Bio-Mass Backpack*, part of the series *Designs for a World of Many Worlds: After the Festival*, 2022-23. Photo: Joshua Riesel.
Dunne & Raby, *Human Bio-Mass Backpack (detail)*, part of the series *Designs for a World of Many Worlds: After the Festival*, 2022-23. Photo: Joshua Riesel.

Viewer as World Maker

Returning to the idea of a design school as a place for nurturing imaginative thought, a place where imagination can thrive, in theory at least, we are still surprised at how difficult it is to cut the ties that link back to the present. To think freely. This is definitely not a criticism of students; we struggle too. It is more of an acknowledgment of just how effectively the design imagination has been constrained.

Stepping outside existing reality and embracing unreality to explore new perspectives is far more difficult than it might first appear. The gravitational pull of the familiar can be overwhelming for those of us trained in design to focus on the here and now, and working within its constraints. We are not advocating that all design make this shift. But we do need to make room for some design to look at these possibilities.

The fear of being unrealistic is powerful. There is an incredible gravitational pull toward realism, or what philosopher Richard Routley has called a "reality fixation." Toward designing for how things already are, or simply reinforcing or extending them. But in order to find new ways of thinking about the world or reality, and our place in it as designers, it is necessary to momentarily suspend the conventions and spend time in parallel realms where different rules and norms prevail. A suspension of not just disbelief but also the real. We are not advocating permanent residency in a fantasyland but instead to make forays into unreality, to see what can be found and brought back. A little like the Stalker in the Strugatsky brothers' novel *Roadside Picnic*.

In an intensive one-week class we teach each winter at The New School, we ask small groups of three to four students to spend some time analyzing a set of specially chosen, decontextualized images, and then to tell the rest of the class about the world they see "evidenced" in those images. What can they tell from (or project onto) the clothes, postures, gestures, facial expressions, furniture, landscapes, and so on, about politics, social norms, family structures, economics, or worldviews. For example, what new thoughts might Olafur Eliasson's *Community Compass* (2017) spark when it comes to thinking about space differently, and its navigation. Even the style of representation, angle of capture, composition, and medium can suggest something about these fictional people and the glimpses into their world we have provided.[19] This is a little like a designerly Rorschach test. But it works. Repeated a few times, students read the images from the perspective of world makers, using them as prompts to imagine other worlds and ways of being. Their curiosity is aroused, and once released, it is difficult to put in back in the bottle. We began to talk about the world hinted at in the picture as "someone else's world"—that is, one different from the one we are collectively working within at that moment. Where things that cannot exist in our world can exist there due to a different set of values, beliefs, and ideally, ontology. Severing the invisible elastic thread that seems to always pull the design imagination back to the here and now. We encourage the students to hold onto that feeling for the week while developing a "scene," or glimpse into an imagined world based on their ongoing research and interests. One of the greatest challenges is resisting the tendency to collapse their ideas into existing reality and align them with how things work now.

To help resist this pressure, we provide guidelines not only for the students but also for ourselves, to prevent us all collectively from slipping back into prevailing norms. For example, shifting from "familiarity to estrangement"; if the scene looks too familiar, then it is not working. But if it is so strange that it appears random, that's not good either. From "transparency to evocation"; it is not about trying to communicate or convey a precise message, which requires established design languages and conventions, but rather about gently jolting the imagination and engaging the viewer, drawing them into the imagined world. And from "convention to invention"; again, resisting the need to make sense in a conventional way, and instead exploring different kinds of logics, associations, and possibilities. These help move beyond the familiar—transmitting a message, solving a problem, demonstrating an idea, raising an issue, or showing a way—to prompting, stimulating, and suggesting. Opening up possibilities rather than prescribing. Searching for meaning rather than solutions, skillfully navigating the edges of weirdness, surreality, absurdism, nonsense, and maybe a little pataphysics too.

This exercise also causes a shift in the way we engage with designed objects. Usually, viewers are encouraged to view them through the lenses of use, technological progress, communication effectiveness, art and design history, designer biography, and so on. It is only when we enter an ethnographic museum that we begin to look at objects in the way we are writing about here. Adopting the mindset of an anthropologist rather than user or design historian. In the book *Extinct: A Compendium of Obsolete Objects*, the authors apply this approach to historical artifacts that are no longer in use, or sometimes, no longer even exist for all sorts of reasons.[20] It is what they tell us about how the world was or where people thought it was heading that is most interesting. Here, we build on this as a generative approach to imagining how things could be otherwise, shifting emphasis from the object itself to the viewer's imagination and willingness to speculate.

The viewer-reader becomes a world maker. Interpreting unfamiliar animal and human figures, furniture, clothing, vehicles, devices, buildings, landscapes, technologies, models, manners, social protocols, drawings in relation to imagined cultures, ontologies, and cosmologies as though they belong to imagined worlds that expand their imaginative horizons. This is about looking at unfamiliar objects and images in order to imagine the nonexistent worlds they could belong in. Reverse engineering objects as an archaeologist might, and letting the object work on the imagination, not just reason.[21]

The hope is that this approach draws out thoughts, ideas, and values, buried deep inside us, letting them find form in apparently random images. Exploring new ways of interpreting objects through a world-making lens that complements more familiar art historical, biographical, and pragmatic approaches to interpretation. Seeking hints rather than complete thoughts and making room in the world for objects that transcend logic-constrained design.

Maiko Takeda, *Atmospheric Reentry* collection, head-covering mask and hooded cape. Photo: Sho Kobayashi for *Disegno* 7 (Autumn-Winter 2014-15).

3967

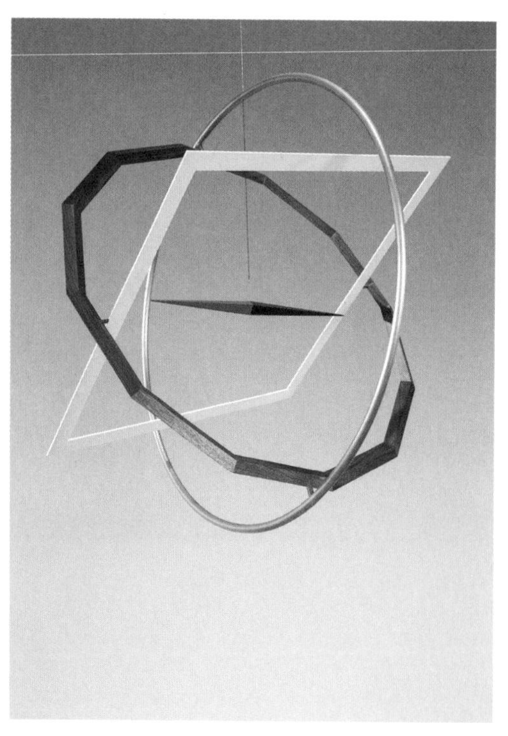

318 taken 1903 Oct 16

Tempest Anderson, *Man Standing in a Spiracle on a Lava Plain*, 1893. Yorkshire Museum (York Museums Trust). Image courtesy of York Museums Trust, public domain.
Olafur Eliasson, *Community Compass*, 2017. Artwork material: stainless steel, wood, paint (brass, white, and gray), and magnet. Linear dimensions: 90 × 100 × 100 cm. Courtesy of the artist, i8 Gallery, Reykjavik. © 2017 Olafur Eliasson.
Alexander Graham Bell facing his wife, Mabel Hubbard Gardiner Bell, who is standing in a tetrahedral kite, Baddeck, Nova Scotia, October 16, 1903. Part of Gilbert H. Grosvenor Collection of Photographs of the Alexander Graham Bell Family (Library of Congress).
Taiyo Onorato and Nico Krebs, *The Hypochondriac*, 2004.
John Seven, *The Love Lost Argument: Snofe Lakes*, 2006.

5 A Publi
Library o

c Lending
f Things

In *The Universe of Things: On Speculative Realism*, Shaviro describes a specific relationship to things as "lures." A relationship that is still a little undervalued in design, but essential if the aim is to engage the mind:

> Rather, we should say that things proposition me or that they offer me a certain "promise of happiness" (to cite Stendhal's famous description of beauty). The qualities of a thing–or more precisely, what [Alfred North] Whitehead calls the "eternal objects" or potentialities that are incarnated in it–are only the bait that the thing holds out to me in order to draw me toward it. It may be that a particular thing dazzles me when it rises up from the depths, or it may be that it intrigues and bemuses me by withdrawing into endless labyrinths. But in either case, a lure has been "proposed for feeling, and when admitted into feeling it constitutes what is felt" (PR, 187). When I respond to a lure–and even if I respond to it negatively, by rejecting it–I am led to envision a possibility, or to "entertain a proposition" (PR, 188), and thereby to feel something that I would not have felt otherwise.[1]

This "feeling something that I would not have felt otherwise" is too easily overlooked in design, where the focus is usually on external rather than internal worlds. Shaviro goes on to write, "This means that as the result of 'entertaining' a lure, I have somehow been transformed–whether grandly or minutely. I have selected one definite outcome from among 'the penumbral welter of alternatives' (PR, 187). As a result, I have become–however slightly or massively–a different entity from the one that I was before this happened. I am no longer the same as I might have been had I not been moved by this particular 'flash of novelty' (PR, 184)."[2]

This is exactly the kind of interaction we are interested in here. One that is more contemplative and gentle, but still impactful, just on our inner rather than external worlds. Designers are trained to focus on the world out there, made from materials, systems, and technologies–stuff as opposed to the worlds we carry around inside us–made from ideas, values, beliefs, and thoughts. Of course the division is not always so clear and there is a good deal of overlap, but where one chooses to focus their efforts can lead to different kinds of design. And maybe, different places of encounter.

There are plenty of places we can visit to learn about how the world lying outside our minds is designed and made, or has been in the past and maybe in the future too. But where can we go to reflect on how the worlds we carry around inside us are formed or even designed, and what they are made of? A place that encourages deep and subtle transformative relationships with designed objects. Something more akin to a library of things. A place where, like libraries of books, people can visit, study, access, and borrow objects. A place that would offer visitors time and space to spend with objects. There would be archives, reading (or looking) rooms, and other spaces that provide more intimate encounters with objects designed for more cerebral uses. The Archive of Impossible Objects from chapter 2 would probably comprise one section in this library, and maybe the Partial Inventory of National Dreams Made Physical from chapter 7 as well.[3]

Objects in a place like this are not destined to be manufactured or even reproduced but instead simply exist. They are theory-objects made from a fusion of matter and ideas, closer to models than prototypes. They sit in the world slightly differently from other designed objects and serve a different purpose, requiring different forms of interaction and engagement when encountered. Rather than telling or even showing us how things should be, they are designed to aid the process of undoing current thought. To produce cracks in the many assumptions we take for granted. Triggering a process of conceptual "taking apart" as Rebecca Soling puts it so well in her book *Eve Said to the Serpent*. Not of things, but of ideas:

> [Thus] we live in a world that first existed inside the heads of others, a world built up through innumerable sustained acts of intentionality, a world where everything speaks not of nature and her processes but of its makers in their resistance to those processes. In a very real sense we can be described as living inside the heads of others, in an excess of interiority that obliterates our own relation to material origins, to biologies, to our bodies. In some way, making was intended to override the givens of nature, to create a world; that world has itself become a given whose terms are more limited in their scope for imagination and act. The world is so thoroughly made it calls for no more making, but for breaching its walls and tracing its processes to their origins. "Taking apart" has become the primary metaphor and backward" the most significant direction: the creative act becomes an unraveling, recouping the old rather than augmenting the new.[4]

Anamorphic Fictions

If the goal is to use fiction to open the mind and encourage imaginative thought, then extrapolative futures might be a little too straightforward; the need to be plausible, rational, or possible ties them too closely to existing realities, of which they are of course a version. Literature offers some alternatives to futures as a primary framing device for fictions concerned with science and technology. Le Guin's "thought experiments," H. G. Wells's notion of "domesticating the impossible hypothesis," and philosopher Quentin Meillassoux's idea of "extro-science fiction" all abruptly take the reader into a parallel realm that may or may not be a version of our own world, where something significant has changed or is different.[5] This approach rarely bothers with the niceties of long-established literary transitions or portals, such as extended sleeps, dreams, hidden doorways, faraway planets, wardrobes, and other thresholds between the shared world of the reader and author and the world of the story.

In design, there is perhaps less need for a portal; viewers are instead simply confronted with a physical fragment from the imagined world, brought into theirs, appearing unannounced. It just is. A piece of real, conceptually elsewhere, but materially present. With this comes a slightly different approach to aesthetics, and rather than aiming to convince, persuade, or trick the viewer (through verisimilitude), it needs to engage them, hold their attention, and send their imaginings off in new directions, or what we called in *Speculative Everything* the "aesthetics of unreality."

Encounters with objects like these do not elicit a leap or jump in the viewer's mind; it is more nuanced, more like a reality shift, a form of anamorphic fiction that necessities a change in perspective. In our own work, we often make use of existing object types from everyday life: they appear familiar, but on closer inspection are not. Looked at from an existing perspective it might not make sense, yet shift (conceptual) perspective slightly, and it does. It is making this shift that matters.

Ideas Made Physical

So how do you approach the design of objects like these, designed not to solve problems but instead spark reflection? What qualities do they need to embody if they are to be used in this way? Designs after all, despite a growing interest in fiction, are still almost always read as something waiting to be realized. Nearly always taken literally. These designs need to have materiality but also maintain fictional status.

In the context of speculative forms of design practice, the physical object has a fluid status. Is it film prop, prototype, model, or something else? If it is a film prop for an imaginary film, or a diegetic prototype as it is also known, then it is the story world that takes priority. The object serves as a gateway or window into the narrative. These kinds of objects need to closely relate to the story if they are to be effective. Designers often use objects to transmit stories, whether a set of brand values or political message. The design serves a didactic purpose and becomes a little reductive here. We prefer to reverse this so that the story world serves as a way of generating an object, where once complete, the narrative falls away like a scaffold, allowing viewers to move off in new directions rather than attempting to resolve a puzzle. Once the object is brought into existence, it is the work it does on the viewer's imagination that matters.

A prototype usually tests an idea against existing realities, whether technical, physical, ergonomic, or so on. The question it answers is always, Does it work? People expect prototypes to be aligned with prevailing realities, to be ready for implementation. This need for realism limits its usefulness in this context. It is probably the model that best describes the kind of objects found in the Public Lending Library of Things. But even models can mean different things.[6] Philosophers of science have written extensively on the scientific model as well as distinctions between theoretical models based on mathematics or statistics and physical models such as a model of DNA. We like this definition: "Scientific models are (functional) mental representations designed by users to represent aspects of the natural world in order to realize certain cognitive or practical goals."[7] But this is still alittle different from our use of models. For us they are a medium.[8] Firmly located in the here and now, physically, but conceptually elsewhere. Not from a future, past, or present; just not here, not now.

As Christian Hubert writes in *The Ruins of Representation*:

> The space of the model lies on the border between representation and actuality. Like the frame of a painting, it demarcates a limit between the work and what lies beyond. And like the frame, the model is neither wholly inside nor wholly outside, neither pure representation nor transcendent object. It claims a certain autonomous objecthood, yet this condition is always incomplete. The model is always a model of. The desire of the model is to act as a simulacrum of another object, as a surrogate which allows for imaginative occupation.[9]

A model, therefore, is never the full story, and like a diagram, is simplified in order to emphasize particular features. Being abstractions, models also offer more potential to explore aesthetics and engage the viewer in a variety of ways. They can straddle different reals, systems of logic, sense and order. Material embodiments of a helpful wrongness.

The aesthetics of objects designed for reflection operate a little differently from other designed objects. They need to hold the viewer's attention, draw them in, and avoid easy resolution through too close an alignment with existing reality; they need to be a little off. With these kinds of designs, there are many other aspects besides meaning and function to consider—structure and composition, scale in relation to the viewer's body (too small and it feels like a toy; too big and it is ungainly), how the eye can be taken for a walk (as opposed to absorbing the object visually in a blink), degrees of abstraction, subtle contradictions, pleasurable wrongness, and lack of detail. Too much detail and it creeps closer to reality; too little and it can appear cartoonlike.

On the surface, models, which are stripped down, abstracted, and devoid of detail, might appear to embrace modernist or at least minimalist aesthetics, but that is not the intention. Their minimalism has more to do with signaling that they are not "actual" products but instead

Bovenbouw Architectuur, *Composite Presence*, 2021. Photo: Filip Dujardin.

theory-objects. If anything, they reference the pared down functionalism of engineering. The objects in this library would be ambiguous, open-ended, and suggestive—designed with the intention of eliciting curiosity as opposed to allaying practical concerns. Rather than aiming for verisimilitude, the emphasis is on showing the seams and joints of a fiction. They need to be interesting instead of believable—both appealing to the imagination and engaging the intellect. This is a very different way of "reading" designed objects that still needs to be encouraged. Interestingly, in *The Architectural Models of Theodore Conrad*, Teresa Fankhänel suggests that the 1939 New York World's Fair "was not just a showcase for the new material and aesthetic world of a modern age in Kodachrome colors, but also the moment a wide audience first became acquainted with architectural models as a medium, which happened concomitantly with the modeling craze which swept across America in the 1930s."[10] It feels as though this public literacy for reading models has long ago been lost; maybe it is time to embrace it again, but in less propagandistic ways, and for objects, not buildings.

Models have two other qualities that are important for the kind of experience we are thinking about here. The first is presence. In a highly mediated world, where the digital image dominates, usually accompanied by a textual explanation or interpretation, it is very special to encounter an actual physical object in a space, unmediated, inviting us to open our imagination to it and listen to what it has to say. The second is scale. While prototypes are nearly always one-to-one, models rarely are. Encountering objects at the wrong scale or size is enjoyable, although there are scales that have themselves become too familiar, such as a doll, toy car, and so on. Scale can be adjusted to feel slightly unusual or unfamiliar for the type of model we are viewing or interacting with. The *Composite Presence* project created by Dirk Sommers of the Antwerp studio Bovenbouw Architectuur for the Belgian Pavilion at the 2021 Venice Architecture Biennale does this nicely.[11] Fifty models of recent buildings from Brussels and Flanders are arranged in a fictional town. At 7:100 scale, they are just big enough to become more than models but less than buildings. It is clear their purpose is something other than simply showing viewers their designs.

Imaginative Mobilities

We had an opportunity to experiment with this in a project called Imaginative Mobilities that we did with colleagues at The New School, supported by the Mellon Foundation, exploring ways of bringing design and the social sciences into conversation around new thinking on borders and mobility.[12] From the proposal:

> Much of the debate on borders—both academically and politically—has revolved around a dichotomy: whether they should be open, or closed. The open borders argument is about free and unfettered movement for all; and the closed borders argument suggests people should be able to create and maintain an inside and an outside. Neither side asks, however, whether we might reconcile borders in different terms—such as permeable, partial, temporary, multilayered—or in different forms, such as welcome lounges, flyways, or weather fronts, shifting hour by hour depending on membership.[13]

The aim was to generate more expansive border imaginaries. Not to design solutions, but design propositions, useful fictions, and hypothetical scenarios in order to facilitate different kinds of conversations across disciplines. To facilitate further thought, discussion, and imagining.

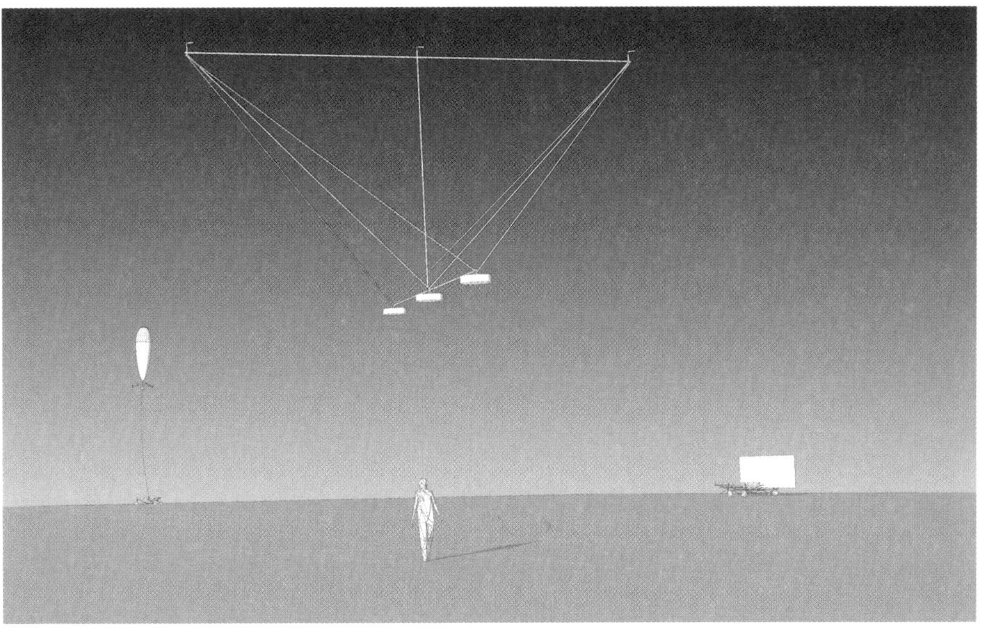

Dunne & Raby, *Transition Zone*, 2018. Illustration: Del Hoyle.

We developed a scenario for a world where borders were no longer viewed as hard lines dividing territories but rather as soft transition zones extending for hundreds of meters between different cultures, ideologies, histories, economic systems, and so on. The focus was on celebrating difference as people moved from one place to another. A series of "drones" and "ground vehicles" would permanently occupy this in-between space, leading people through it. Broadcasting stories, histories, and other content representing the worldviews and narratives that meet there as they draw people through the space. They would be assembled from equipment typically used by security forces to control and often intimidate citizens. Repurposed for cultural ends. Here, they are used to draw people together. Reflected in the appearance of the devices. Delicate, wooden, slightly ungainly structures at odds with the slick, intentionally intimidating technological materiality and structures typically used in drones and autonomous vehicles. More like Alexander Graham Bell's early kite experiments, which looked like they didn't belong in the sky.

Sounds transmitted from local radio stations in regions across each country, located along virtual axes extending far beyond the zone, are whispered from sound canons usually used for crowd dispersal, providing acoustic slices of local cultures in the form of accent, news events, language, vocabulary, and music. Small groups of people wander behind these floating sound spaces enjoying the fusion and hybridization of national stories and histories. Nearby, strange-looking vehicles composed of ad hoc seating and projection screens, or stages for gently amplified conversations between historians and citizens, slowly move through the zone. In the distance, inflatable representations of disputed figures and icons float, present but not permanent, demoted to iconic placeholders for contested stories.

The design proposals took the form of scale models. Not life-size, but still large. Their structures were schematic and lacked detail; the joints were intentionally irregular, each one unique. Not very efficient, but they drew the eye into the object. In the context of this library, they would be available to borrow. In the home, they would sit alongside tables, cabinets, chairs, and lamps, forming domestic tableaux or miniature scenes depicting imagined worlds entangled with the furnishings of exiting reality, meeting more contemplative needs.

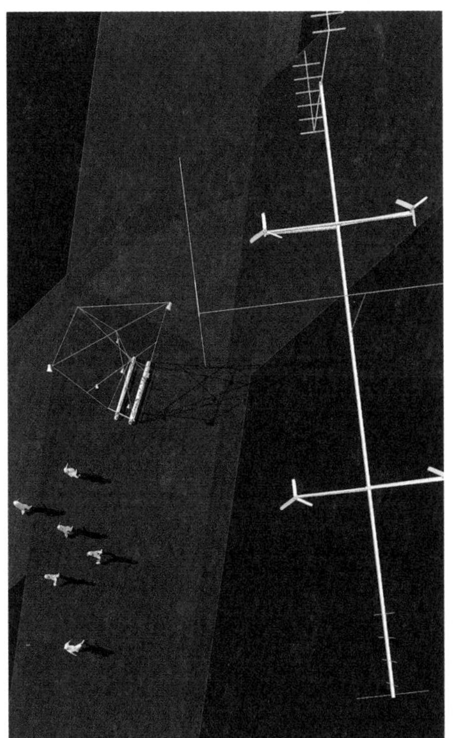

Transition Zone Vehicles, 2018. Illustration: Del Hoyle.

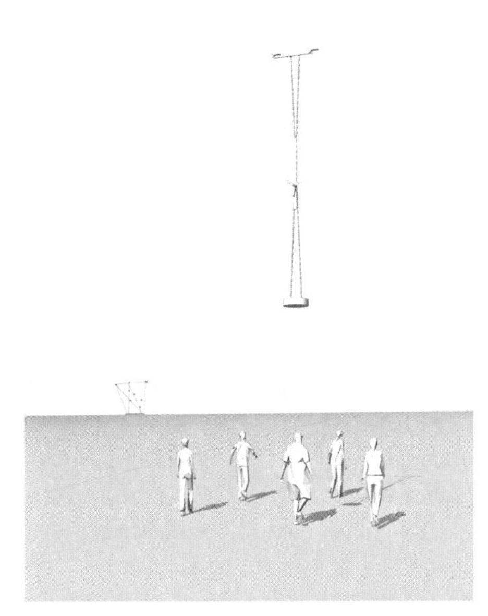

Transition Zone Vehicles: Drone, 2018. Illustration: Del Hoyle.

Alexander Graham Bell (*right*) and his assistants observing the progress of one of his tetrahedral kites, July 7, 1908, part of Gilbert H. Grosvenor Collection of Photographs of the Alexander Graham Bell Family (Library of Congress).

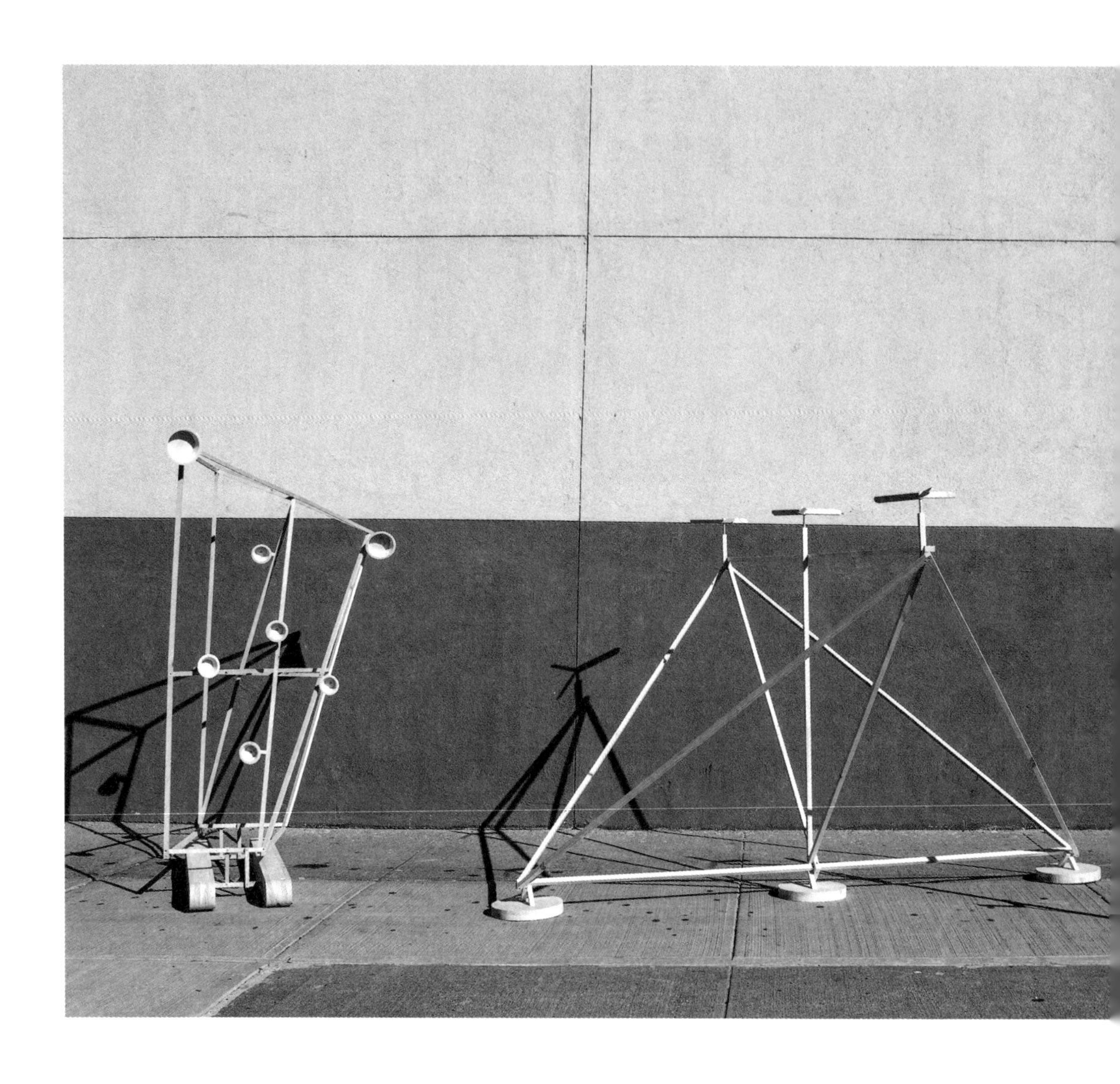

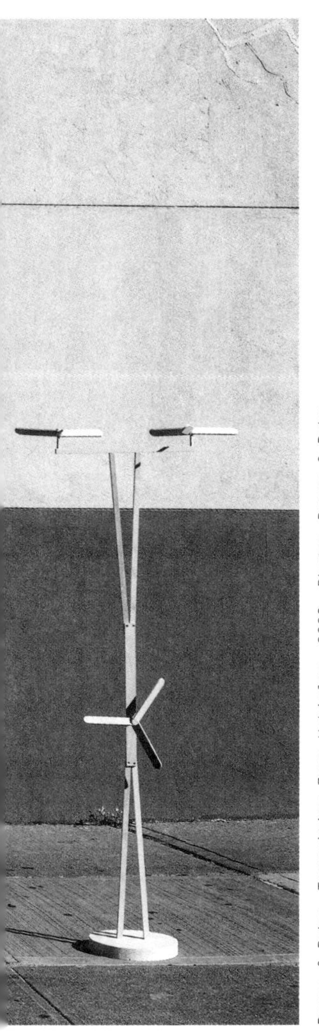

Dunne & Raby, *Transition Zone Vehicles*, 2020. Photo: Dunne & Raby.

Dunne & Raby, *Transition Zone Vehicles*, 2020. Photo: Dunne & Raby.

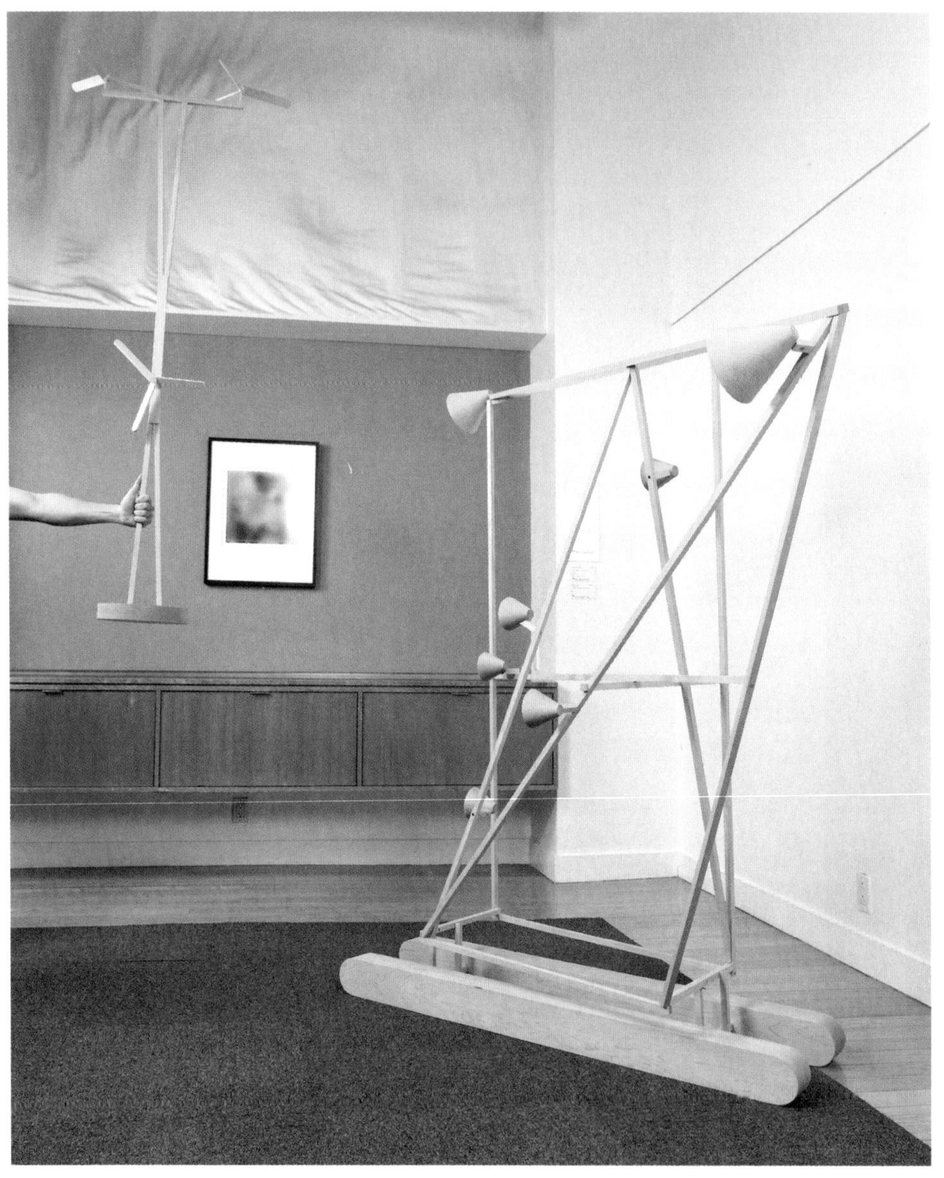

Dunne & Raby, *Transition Zone Vehicles*, 2020. Photo: Yuhan Pan with Yoshe Li and Billy Chen.

Dunne & Raby, *Transition Zone Vehicles*, 2020. Photo: Yuhan Pan with Yoshe Li and Billy Chen.

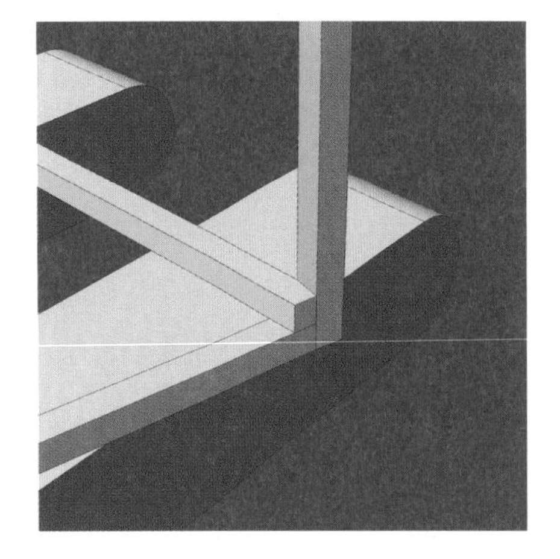

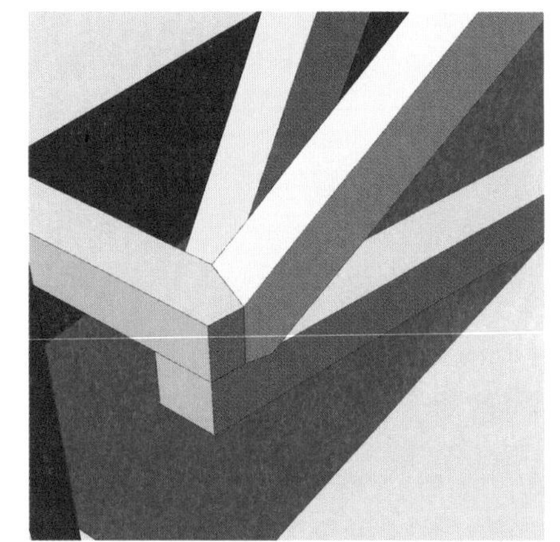

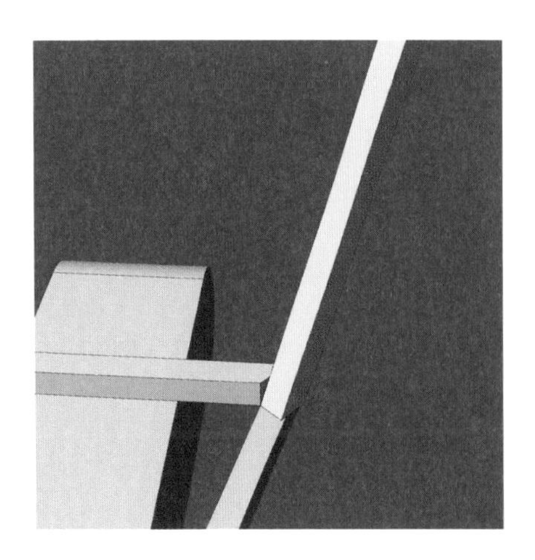

Dunne & Raby, *Transition Zone Vehicles*, 2020. Details. Illustrations: Del Hoyle.

Scientists Studying Moon Phases on Models in Preparatio for US Manned Flight to Moon, 1961. © Fritz Goro, Life Picture Collection, Shutterstock.

Different Shaped Worlds

As models go, globes are quite special. Usually found in isolation, there is always, and can only be, one. The right one. The latest, most correct, and accurate one. It is strangely unsettling to see more than one globe together. A cluster of globes feels odd—too much conceptual redundancy perhaps. The oldest-known (Western) globe to represent the "new world," made from two halves of an ostrich egg, dates to 1504. We have long since passed the point where ambiguous regions are updated on each new globe, unless talking about ocean depths, so there is no modern equivalent of the warning "Hic Sunt Dracones" or "Here Be Dragons" found on the Hunt-Lenox globe (about 1508—one of the earliest-surviving Western globes of earth), which coincidently, is housed in the Rare Book Division of the New York Public Library.

Globes have long been part of the domestic environment, usually showing the earth or sometimes the moon. One of the few industrially produced objects that serves a more symbolic function, meeting different needs from those provided for by its industrially produced domestic kin. It arouses curiosity, allowing journeys to be mapped, stories to be told, and adventures to be planned or recounted, not in the abstract, but situated in the world. Although often associated with home decor and potentially a little kitschy, there have been several interesting recent design interpretations. Oscar Lhermitte, Peter Krige, and Alex Du Preez's topographically accurate *Moon Orbit Lamp* models the moon's surface in ways reminiscent of James Nasmyth's *Group of Lunar Mountains: Ideal Lunar Landscape* from 1870.[14] And Yuri Suzuki's *The Sound of the Earth* consists of a vinyl record of field recordings of anthems and folk music from around the world in the form of a sphere.

For an exhibition marking hundred years of the Bauhaus, *Alternatives for Living. Blueprints for Haus Lange Haus Esters* set in two houses designed in the 1920s by Ludwig Mies van der Rohe in Krefeld, Germany, we were asked to design an everyday object that suggested another possibility

The New World Depicted on the da Vinci Globe Dating from 1504. © Stefaan Missinne, 2013.

for daily life. We chose a globe, something that allowed us to introduce other ways of viewing and making sense of the world(s) in a domestic setting. Not one globe, but a collection of ten. It includes worlds existing at the edge of the known universe only recently captured in image form. Worlds from the imagination of writers that borrow concepts from the fringes of mathematics and physics. The imaginings of conspiracy theorists who, suspicious of science, attempt to offer not very credible alternatives to conventional thought. Thought experiments by curious minds that wonder how earth might work if it were a different shape. And one that despite its small scale and mundane presence, references some of the most complex puzzles facing scientists today: that this place we find ourselves inhabiting might possibly be just one of many universes, some of which may even be parallel versions of our own.

The Asian Continent Based on Ptolemy on the da Vinci Globe Dating from 1504. © Stefaan Missinne, 2013.

Each globe therefore suggests a different way of thinking about the idea of a world and reality itself. They represent different kinds of imaginations. Arranged in a neat sequence, they would remain in their own little zones, isolated. Instead, the collection is designed to place each one in close proximity to another, all coexisting on a domestic-sized tabletop, jostling for room within one space. The room we exhibited in was historically the "women's room," and we like to think that less constrained than their menfolk, the women contemplated how the world might be otherwise using a collection of globes like those described here. Since then, each time they have been exhibited, curators have used tables rather than plinths to display them.

Oscar Lhermitte, Peter Krige, and Alex Du Preez, *MOON*, 2016.

James Nasmyth, *Ideal Lunar Landscape*, ca. 1873.

Yuri Suzuki, *The Sound of The Earth*, 2009–2012. Photo: Hitomi Kai Yoda.

Dunne & Raby, *Archive of Impossible Objects: Globes*, 2019.

Dunne & Raby, *Archive of Impossible Objects: Globes*. Exhibited at Haus Lange Haus Esters, Krefeld, Germany, 2019. Photo: Dirk Rose.

Dunne & Raby, *Archive of Impossible Objects: Globes*. Exhibited at Mudac, Lausanne, Switzerland, 2023. Photo © Etienne Malapert and legend: View at Terra, Space Season, mudac 2023.

Dunne & Raby, *Archive of Impossible Objects: Globes*, detail. Exhibited at Mudac, Lausanne, Switzerland, 2023. Photo © Etienne Malapert and legend: View at Terra, Space Season, mudac 2023.

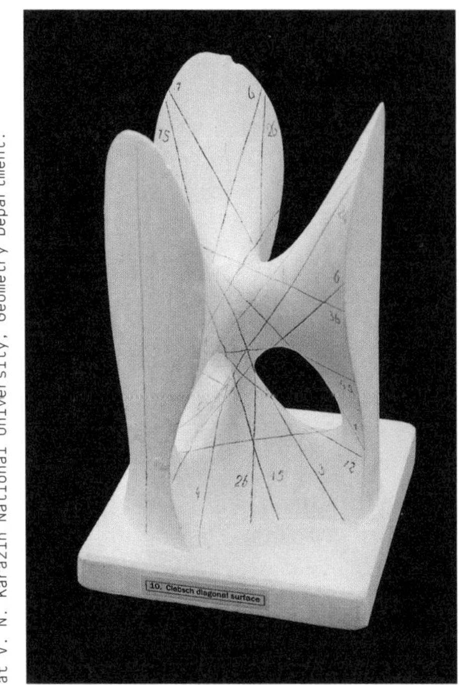

Model of Clebsch surface. The model is in the Geometrical Collection at V. N. Karazin National University, Geometry Department.

The globes are white, mathematical-model white, reflecting a long tradition of scientific and mathematical models made from plaster, many of which were made in Germany. They are accompanied by three topographical maps intended to prompt further imagining about the workings of each world. Each one makes use of conventional markings, but in ways that do not quite make sense. They suggest speculative atmospheres, climates, geologies, forces, and fields different from those shaping our world. Besides existing in physical space and the imaginations of visitors to the exhibition, they are part of an informal online library of imagined worlds distributed across the internet and accessed through search engines.

At this point, they reflect Western ideas, but in future iterations they could include models inspired by Egyptian mathematician Mostafa A. Abdelkader's calculations for an inside-out cosmos. Or the world at the center of *The Classic of Mountains and Seas*, a guidebook to a mythical world of ancient China from the third century BC to the second century AD. Or representations of the cosmos as a lidded vessel reflecting the Yoruba peoples' cosmology in Africa. Elements of Aboriginal cosmology where the earth is a disk and the sky is a dome. As well as cosmological models based on Vedic Hindu and Tibetan belief systems.

There is one globe in the set that aligns with how things are now. A perfectly spherical model of earth. There are no markings on our model earth, but if there were, they would not show how it is now or has been in the past but instead how it might look in a hundred to two hundred years' time based on current predictions for how the earth will be remodeled due to climate change. Maybe the surface of this globe, if it were to be mapped, would once again have zones declaring "here are dragons," where dragons now stand for unimaginable environmental horrors and transformations potentially on their way, unless we act soon.

A cuboid earth might seem a step too far in thinking about alternative-shaped planets. After all, a sphere seems the most natural shape for a planet considering that anything from four to six hundred kilometers in diameter, depending on the material, will end up as a sphere due to gravity, at least in our universe. But surprisingly, cuboid planets do have a history—that is, imagined ones. A popular online science series tells the story of a man named only as Arndt who in 1884, claimed to have discovered a cubic planet lying beyond Neptune. In an article titled "The Cubical Planet" in the *New York Times* of November 16 of the same year, physicist Theodore Vankirk explains how such a planet might work, focusing on how gravity would feel different on a cube versus a sphere.[15] Each face would feel like a bowl as one would only be able to stand upright at the center of a face. Walking toward an edge would feel as if you were going uphill. Later, in the 1950s, Scotlund Leland Moore claimed to have discovered a cubical planet he called Aocicinori, and that he could communicate telepathically with its inhabitants, whom he illustrated, as well as maps. A mini cult sprung up around the planet.[16]

More recently, when she was an astronomer at Cornell University, Karen Masters also explored the practicalities of how a cubic earth might work. Her reflections on a personal blog were sparked by a question from a secondary school student interested in how the earth's weather would change if it were in the shape of a cube.[17] She points out that oceans would probably be located at the center of each face and try to form a sphere due to gravity; likewise, the atmosphere would form a sphere with the corners of the planet poking through, forming "mountains." Life would be limited to a narrow band around each sea, and gravity would be strongest at the center of each face. Although cubical planets do not exist, or at least, no actual examples have been sighted yet, they do seem to have place in the imagination, where they serve as models for experiments in how gravity, for example, might work differently, giving rise to a different geography and weather.

Another planetary modeler has focused on what a toroidal version of earth might be like. In 2014, scientist Anders Sandberg developed a highly detailed thought experiment using computer simulations to explore properties such as gravity, light, geosphere, atmosphere, hydrosphere, biosphere, moons, and tidal forces. Referencing work by mathematicians Henri Poincaré and Sophie Kowalewsky, as well as astronomer Frank Watson Dyson, he concludes, "It looks like a toroid planet is not forbidden by the laws of physics. It is just darn unlikely to ever form naturally, and likely will go unstable over geological timescales because of outside disturbances." One of his more interesting observations was how days and seasons would work depending on tilt, such as at zero tilt,

Dunne & Raby, *Toroid Planet Topographic Map*, 2019, part of *Archive of Impossible Objects: Globes*. Illustration: Carolyn Kirschner.

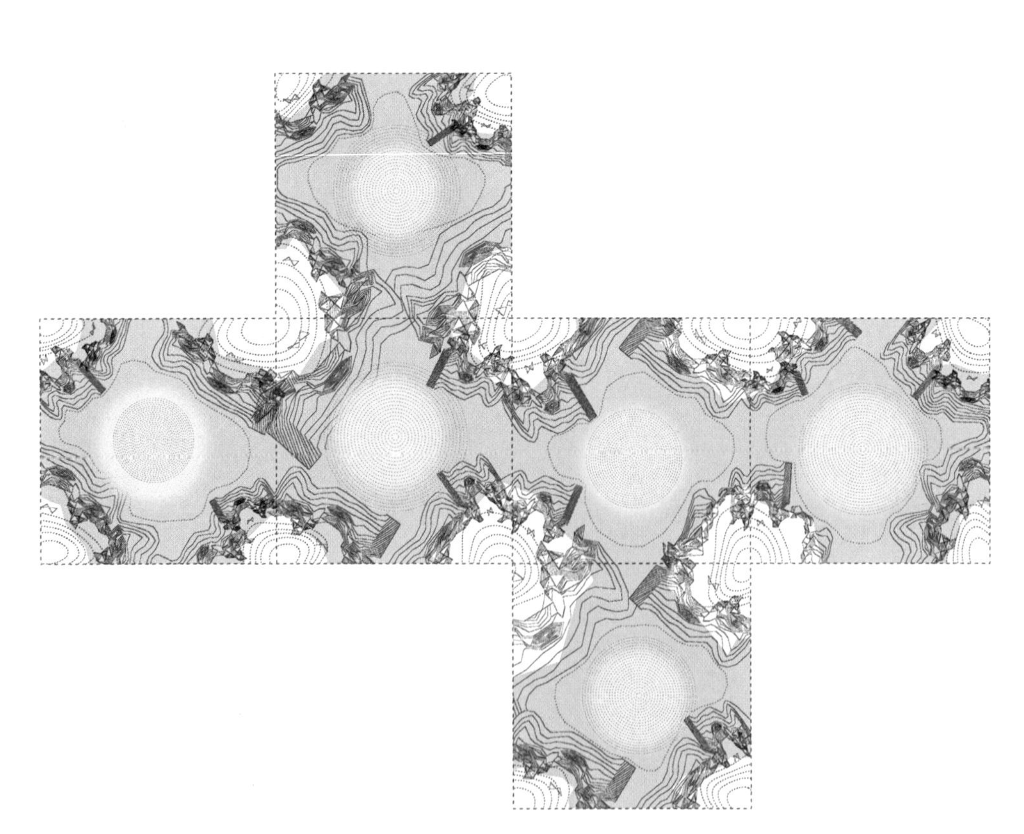

Dunne & Raby, *Cuboid Planet Topographic Map*, 2019, part of *Archive of Impossible Objects: Globes*. Detail.

Dunne & Raby, *Möbius Planet Topographic Map*, 2019, part of *Archive of Impossible Objects: Globes*. Illustration Carolyn Kirschner.

> the hubward side will never get any sunlight: the sun is always hidden below the horizon or by the arc of the world. At the poles the sun is moving just along the horizon, and slightly inwards there will be a perennial dawn/dusk. The temperature difference will be big, with the interior at subarctic temperatures: this is not entirely different from a tidally locked world, and we should expect water (and maybe carbon dioxide) to condense permanently here. The end result would be an arid (but perhaps not super-hot) outer equator, possibly habitable twilight polar regions, and an iced-over interior.[18]

Several fiction writers, unconstrained by existing planets, have unsurprisingly gone further. Starting not with alternative forms for existing planets but rather alternative physics that give rise to new planetary forms. In this kind of science fiction, all too often the settings might be exotic, but the social aspects are not. The politics and economics of these imagined worlds are simply a cut-and-paste version of those on earth. When the physics of an unusually well-designed planet are foregrounded, familiar economic and social systems seem even less believable, although for most readers of fiction like this, it is probably the design of the world that provides most pleasure.

Greg Egan's hyperboloid planet, the setting for his novel *Dichronauts*, is a good example. Here, space and time each have two dimensions rather than three and one, as our space-time does.[19] As Egan writes on a website explaining the physics behind his world,

> In the world of Dichronauts, a vertical object that topples over in the wrong direction is in big trouble. If the gravitational up/down direction is "space-like," then there will be both "space-like" and "time-like" horizontal directions, and if it's possible to fall in a "space-like" direction then that will be no different from toppling over in our own world. That's the safe way to fall. But we are interested in what happens if—for whatever reason—an object topples over in a direction of the opposite kind to up/down.[20]

As hard as we try, we still cannot imagine what this might look like, but this for us its value: to render something like space-time strange, to defamiliarize it in ways that force us to consider our own space-time as one of many possibilities. Dichronauts is admirably complex and mind stretching, but sadly, the characters and their motivations feel far too terrestrial for such a beautifully crafted mathematical construction.

Christopher Priest's *Inverted World* (1974), which opens with the now-famous line "I had reached the age of six hundred and fifty miles," immediately establishes that this is a very different world from our own. The cosmology of the world's people is also imaginative and intriguing. Its social aspects emerge from the unique qualities of the world and its strange physics too. Occasionally, the reader catches glimpses of its physics, and the gravitational distortion of space and time it produces. For example, as you travel north of the city, which is slowly crawling along the surface of the planet on a set of rails that have to continuously be moved from the rear to the front of the city, objects appear infinitely tall and infinitely thin, and time speeds up. And when you travel south of the city, people and objects gradually become infinitely short and infinitely wide, and time slows down. The goal is to reach the "optimum" for the city, where the slope of the curve is forty-five degrees from the horizontal, and time and space are as they should be.[21] Although never stated in the book, several readers have suggested that the planet, which could be earth, has taken on the form of a pseudosphere due a massive science experiment gone catastrophically wrong.

Like the hyperboloid and pseudosphere, *A Mission of Gravity* (1953) also takes place in a world that models an alternative physics without breaking any fundamental laws to produce its strangeness. Gravity on Mesklin, where the novel is set, is three times stronger than on earth at the equator and 275-700 times stronger at the poles depending on how it is calculated. Its equatorial diameter is 48,000 miles; pole to pole, it is 19,740 miles, giving the planet an oblate form; a day is 17.75 minutes long; and the average temperature is -170°C. The planet has life-forms as well, but they are not so imaginative, and the plot is a little disappointing too. It is most interesting when it explores the consequences of its physics on the inhabitants of the planet:

> Due to the planet's intense gravity, the density of Mesklin's atmosphere varies so strongly with altitude that refraction makes it look bowl-shaped. The Mesklinites can see that the world curves up around them, so they believe that they live in a giant bowl. They are skilled sailors and map-makers and should know better, however when you are measuring distances on a curved surface, there are two different shapes that will make all the math work out (convex and concave). The Mesklinites chose the wrong one for their maps and never noticed. The result is perfectly accurate and usable maps based on a fundamentally flawed premise.[22]

But even these imagined planets can seem a little tame compared to exoplanets located outside the solar system whose environments often sound far stranger than fictional planets, although they still conform to the laws of physics. For example, Corot-7b, where it rains rocks into lakes of molten lava, or HD 189773-b, where glass winds can blow at speeds of 5405 miles per hour, paling in insignificance against those of planet HD 189733-b, where the winds can reach 21,747 miles per hour. Then there is Kepler-16b, which has two suns, and HD 131399Ab, which revolves around three suns at once, rotating so slowly around its suns that one of its years takes 550 of ours and the sun only sets every 275 years. Or GJ 1214b, where water occupies 10 percent of the planet's total mass, meaning it is covered in an ocean hundreds of kilometers deep compared to earth's, which is just less than 11 kilometers deep. And KELT-9b, where temperatures are about 7800°F/4327°C due to its proximity to a gigantic star twice as large as our sun.

And on January 1, 2019, Arrokoth (2014 MU69), an object about 20 miles long made of a flat sphere-like part attached to a smaller spherical element, was designated the most distant object ever to be explored by a space probe—in this case, by New Horizons at 4 billion miles from earth, literally located at the edge of the known world. Only recently captured in image form, it is forcing scientists to question how planets form as they journey from being a "thing" in the world to becoming an "object" of human knowledge.[23]

There are, of course, also models of the world that do not conform to science, math, or the laws of physics but instead to a worldview or point of view. We have included a flat earth in our collection of globes because we believe that even if scientifically incorrect or pseudoscience, when not taken literally, it can serve as a reminder that people do in fact live in very different worlds, and this can have real consequences if not acknowledged.

The last of the globes, a modest object, sets up a contrast with the speculative model it represents: a multiverse. The idea that one would even try to make a globe of a multiverse is a bit absurd. It is certainly one of the more difficult concepts being wrestled with today. Even among experts, there is little consensus on what a multiverse might be. Some kind of hypothetical group of universes containing "parallel universes," "parallel worlds," "quantum universes," "parallel realities," or "alternate universes."[24] Beyond the observable universe, there are many others, some with laws radically different from our own, and that means our space-time could turn out to be just one possibility among many. But whether or not this turns out to be true, we believe it is important that ideas like this are brought into contact with the everyday, to sit alongside a vase of flowers, newspaper, or set of keys. To close the gap a little, between ideas that stretch the imagination and daily life by entangling them with the everyday objects.

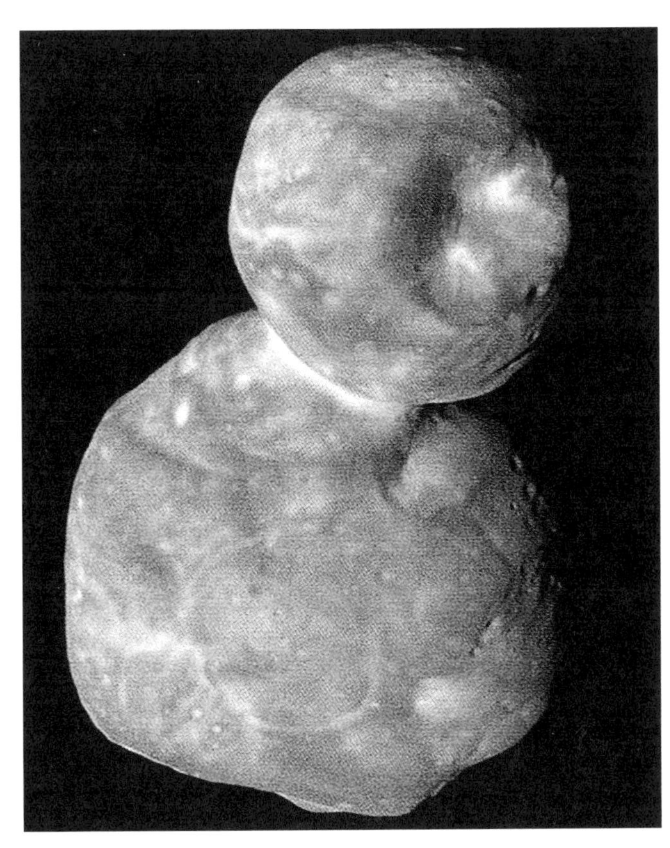

Arrokoth (2014 MU69). NASA / Johns Hopkins Applied Physics Laboratory / Southwest Research Institute, National Optical Astronomy Observatory.

6 The Uni
Kingdoms
a Travele

ted Micro
(UMK),
rs' Tale

Chapter 6

The United Micro Kingdoms (UMK), A Travelers' Tale

The following text consists of extracts from a journal summarizing observations and thoughts from seven days spent in the UMK.[1] During this period we observed its culture, nature, language, mythologies, and outlook, focusing on what is particular to individual micro kingdoms rather than making comparisons across each.

We would like to thank the European Institute for Experimental Nation Building, which generously supported this research; the School of Constructed Realities, which granted us a sabbatical; and of course, all the individuals and groups that kindly opened their micro kingdoms to us during our travels.

Before we deal with our experiences, we should say a little bit about the UMK and how it came to be. Although now regarded as a bold and visionary experiment well on its way to yielding a positive path forward, one cannot underestimate the desperation that led to its formation. There was a time when the United Kingdom seemed on the verge of a complete social and economic collapse; rather than waiting for the inevitable, the nation decided to force a crisis later termed the "preapocalyptic option." A group was formed to develop four prototype states to discover which, if any, might provide hints for a model able to function and thrive within a new twenty-first-century world order.

The UMK was divided into four zones, each driven by a group promoting specific technological and political agendas. Briefly:

DIGITARIANS depend on digital technology and all of its implicit totalitarianism—tagging, metrics, total surveillance, tracking, data logging, and 100 percent transparency. Their society is organized entirely by market forces; citizen and consumer are the same. For them, nature is there to be used as necessary. They are governed by technocrats, or algorithms—no one is entirely sure or even cares, as long as everything runs smoothly and people are presented with choices, even if illusionary.

Dunne & Raby, *Here-and-Nowists Protesting in Digiland*, 2014, part of the series *UMK: Lives and Landscapes*. Illustration: Miguel Angel Valdivia.

The COMMUNO-NUCLEARIST society is a no-growth, limited population experiment. Using nuclear power to deliver near limitless energy, the state provides everything needed for people's continued survival. Although they are energy rich, it comes at a price: no one wants to live near them. Under constant threat of attack or accident, they live on a continually moving, 1.8 mile (3 kilometers) long, nuclear-powered mobile landscape. Consequently, they are organized as a highly disciplined mobile microstate. Fully centralized, everything is planned and regulated. People are voluntary prisoners of pleasure, free from the pressures of daily survival, communists sharing in luxury not poverty. Like a popular nightclub, there is a one-out, one-in policy, but for life.

Dunne & Raby, *Communo-Nuclearist Control Room, Viewing Platform and Library*, 2014, part of the series *UMK: Lives and Landscapes*. Illustration: Miguel Angel Valdivia.

BIOLIBERALS are social democrats who embrace biotechnology and the new values this entails. They live in a world where the hype of synthetic biology has come true and delivered on its promise: a society in symbiosis with the natural world. Biology is at the center of their worldview, leading to a radically different technological landscape to that of the United Kingdom. Nature is enhanced to meet growing human needs, but people also adjust their needs to match the available resources. Each person produces their own energy according to their needs. Bioliberals are essentially farmers, chemists, and gardeners. Not just of plants and food, but of products. Gardens, pharmacies, and farms replace factories and workshops.

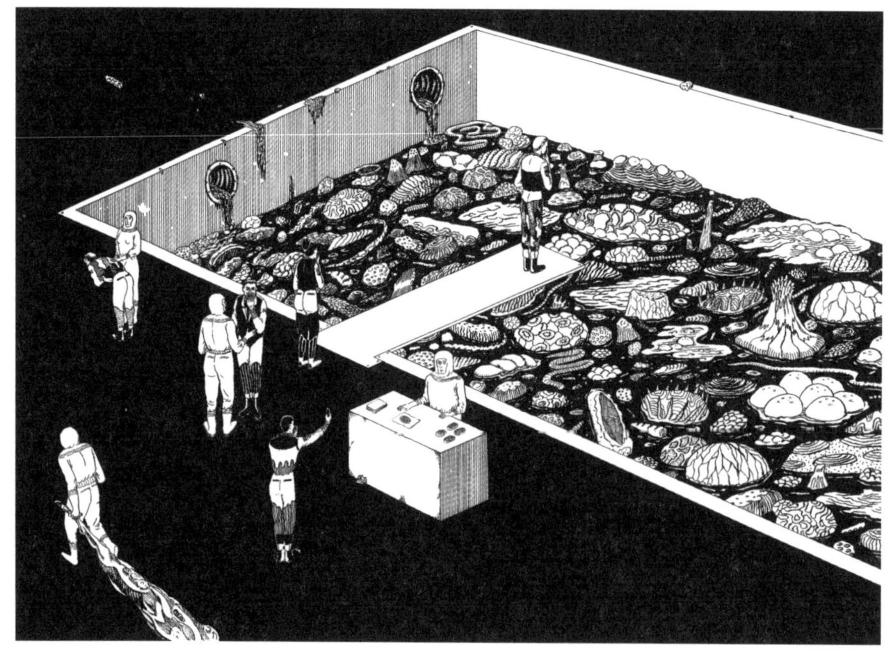

Dunne & Raby, *Digitarians Visiting One of Bioland's More Extreme Attractions*, 2014, part of the series *UMK: Lives and Landscapes*. Illustration: Miguel Argel Valdivia.

The ANARCHO-EVOLUTIONISTS abandon most technologies, or at least stop developing them, and concentrate instead on using science to maximize their own physical capabilities through training, DIY biohacking, and self-experimentation. They believe that humans should modify themselves to exist within the limits of the planet rather than modifying the planet to meet their ever-growing needs. There are a high number of posthumanists among the anarcho-evolutionists–individuals whose physiologies have been improved beyond what is considered naturally human. They essentially take evolution into their own hands. Little is regulated; citizens can do as they please as long as it doesn't harm anyone else. They are the most extreme.

Dunne & Raby, *Bioliberals Visiting an Anarcho-Evolutionist Zoo Lab*, 2014, part of the series *UMK: Lives and Landscapes*. Illustration: Miguel Angel Valdivia.

DAY 1, ARRIVAL: BUREAU OF DISEMBARKATION

Entry into the UMK is always difficult. On this occasion we entered through Digiland, the only part of the UMK linked to other digitally enabled countries. Being researchers, we were able to use the "exceptionally curious guest" visa, which allowed us to stay for seven days. It's a lovely document combining hard information with Escher-like graphic design. Only five hundred are issued in any year, and it took two years to gain ours.

Although we had heard much about it, we were still unprepared for the digitarian landscape. A vast plane of tarmac extending as far as the eye can see, stunning in its confident single-mindedness. There is no doubt here about what they value. Our guide told us the tarmac runs right up to the cliff edges; in fact there are some rather strange markings that we later discovered indicated how much land was being eaten away by the sea. The lack of physical barriers is striking. But self-drive cars, or digicars as they call them, don't really need physical fences because they are geofenced. Rather touchingly, road graphics are still used in extreme situations for the benefit of humans, more as placebo markings than anything else; faith in technology is still not enough when traveling at forty kilometers per hour one meter from a cliff edge. There's something quite beautiful about the monochromatic meeting of digiland's dense black plane and the gray wintery sea. And it's one place you can see curves, along the cliff edge—lots of them.

DAY 1, THE BUREAU OF DIGITARIAN ESCHATOLOGY

Once through immigration, we were immediately taken to the Bureau of Digitarian Eschatology.

Digitarians focus on the short term and how to extract the most they can from finite resources. They have adopted a fatalist position, believing there is nothing humans can do to save the planet and ultimately themselves.

The Bureau of Digitarian Eschatology grew out of a major research project combining big data and predictive algorithms to analyze data from climatology, population growth, food production, land usage, and so on. Its goal was to calculate the end of their world as accurately as possible. Resources are limited, and the digitarians would like to use them effectively; this is their way of dealing with the end as humanely as possible. Once this date was fixed, they identified a second date after which no children were to be born–it would be too cruel to bring them into a dying world–and this additional date was termed -75 Day, a momentous date for all digitarians.

DAY 1, FINDING SOME SCHOOLBOOKS

Later, in our host's home, we noticed some of their children's schoolbooks. Digitarians place great importance on precision, especially with language. Colors, for example, are defined by their spectral characteristics, and much time has been spent defining borders between colors in precise frequencies. Traditional colors are still recognized, but more precisely defined. For instance, red is 740-625 nanometers or 405-479 terahertz, based on the RGB [red, green, blue] standard. There are even impossible colors that exist theoretically but beyond the range of electromagnetic frequencies the human eye can see. Chromatology is taught at school, and every child is familiar with chromatropes, chromoptometers, colorimeters, cyanometers, indigometers, and spectrographs.

Terms for indicating periods of time, such as "in a moment," "soon," "shortly," and "later," are also defined precisely. Apparently this precision is derived from a pre-UMK American National Intelligence Estimates system that attempted to map verbal expressions of "probability" onto percentage values.

For such a pedantic society, the digitarians are constantly wrestling with conceptual and even metaphysical issues. For example, while at the bureau earlier in the day, someone told us how digitarians typically struggle with holistic thought and individual features are more highly valued than a whole. Everything from devices to services and even people are presented as accumulations of features.

At some point early in digitarian history things began to dissolve into lists of features to the extent that it became impossible for them to see *anything* as a whole. Once things began to blur, identities were lost and object typologies dissolved. Language seemed to detach itself from the material world. Various attempts were made by marginal groups such as the thingists to reclaim objects as distinct linguistic entities, and occasionally, under the right circumstances, the continuous mesh of features and possibilities resolved itself into discreet objects, things, roles, and people. But it never lasted.

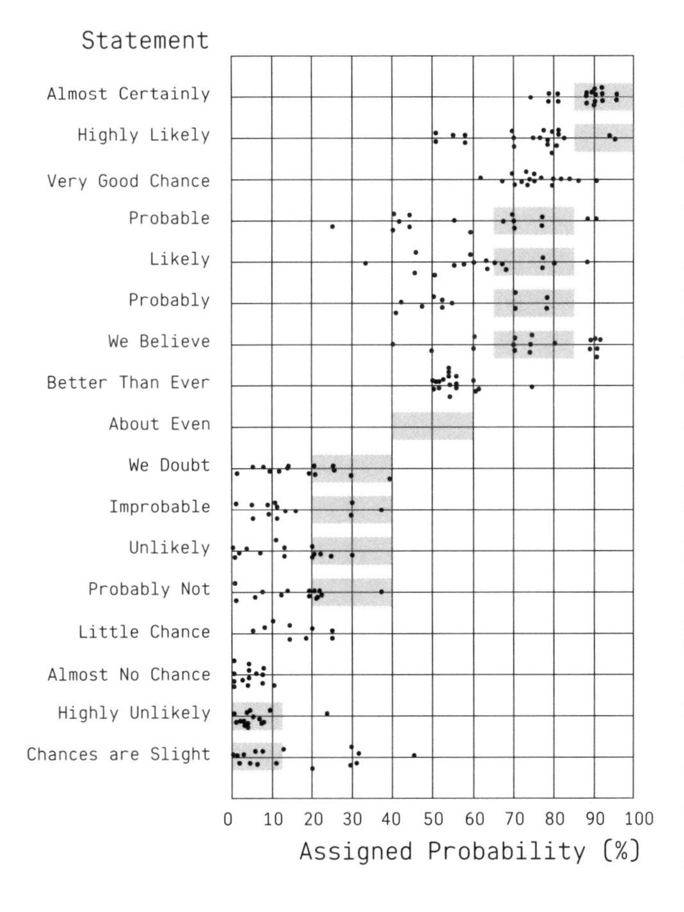

Measuring Perceptions of Uncertainty. Redrawn from Richards J. Heuer Jr., *Psychology of Intelligence Analysis* (Langley, VA: Center for the Study of Intelligence, Central Intelligence Agency, 1999).

DAY 2, A VISIT TO BIOLAND

Digiland was a good base for us; we had common roots. In some ways it was the closest to how the United Kingdom might have been if the preapocalyptic option had not happened. Some of the other micro kingdoms are definitely a little strange—fine to visit, but not the sort of places where we wanted to spend too much time.

For our first trip out of Digiland we traveled to the Land of Beautiful Rotting, a name we gave it after learning that the aesthetics of decay is a critically important concept for the bioliberal community. Approaching the boundary, you start to smell it: a miasma of decaying meat, sulfur, and compost, amalgamated yet piercingly intense, the result of much experimentation and deliberation. People often plan their visit on the strength and direction of the wind, approaching leeward if possible, and entering only as far as necessary.

Before seeing Bioland, if someone had mentioned a biotechnological landscape engineered to facilitate symbiosis, we would have thought of a slightly exaggerated version of our own countryside: oversize plants, abundant foliage, and vibrant colors, all a little cartoonish, a sort of technicolored land of plenty. But Bioland couldn't be more different.

Although organic, this region is neither beautiful nor "natural" looking; whole areas of landscape consist of complex knots of tubes, bladders, and pools—as though some unimaginably large animal had been eviscerated, its innards redistributed over the countryside in dramatic, multicolored pools of harsh chemical hues reminiscent of nickel tailings, although highly ecological. Nothing is toxic; everything feeds into everything else, nourishing, transforming, growing, and mutating. It's wild–inspiring–and sublime even. It is apparent that the inhabitants of the Land of Beautiful Rotting struggled to let go of visions of artificially sustained nature that satisfy the ideals of the past, but once discarded, they began to create a true landscape of the now. This is no accidental Anthropocene.

Bioliberals regard the use of huge amounts of energy to overcome gravity and wind resistance to be counterproductive and primitive. Faster is no longer better. People travel in extremely light, organically grown vehicles, each customized to its owner's dimensions and needs. The bioliberal car combines two technologies: anaerobic digesters that produce gas, and fuel cells that use the gas to produce electricity. Sacs of uncompressed gas cannot compete with the efficiency of fossil fuels–a fuel based on millions of years of preparation compared to one that takes days. The resulting cars are bulky, messy, smelly, and made of artificial lab-grown skin, bone, and muscle, not literally, but in abstracted forms. Wheels, for example, are powered individually using jellylike artificial muscles. Bioliberals work closely with natural forces rather than attempting to escape or overcome them.

DAY 3, ACCOMPANYING BIOLIBERALS ON A VISIT TO THE ANARCHO-EVOLUTIONISTS

Observing our keen interest in dilemmas, a group of bioliberals invited us to accompany them on one of their regular trips to the anarcho-evolutionists.

The next day we headed into the wilds of the anarcho-evolutionist's micro kingdom. There's a long history of trade, exchange, and collaboration between their kingdoms. Anarcho-evolutionists fully embrace self-experimentation. Their outlook is driven by the belief that people should use technology to modify themselves to fit within the limits of the planet rather than using it to reengineer the planet to meet the ever-expanding needs of humans.

By providing advanced medical field facilities for the anarcho-evolutionists, bioliberals gain access to outcomes from experiments they could never dream of, while anarcho-evolutionists can fast-forward their evolution through surgically enhanced modifications. Although it seems like a near-perfect relationship, there is constant debate in the bioliberal world about the ethics of this arrangement. They gain invaluable insights, but also contribute to suffering when experiments go wrong. Our guide was quite troubled by this, and told us he had spent many hours pondering whether bioliberal ethics should be portable and applied wherever they went, rather than their current view, where ethics were somehow attached to the land or state, staying behind when they traveled.

Once they left their micro kingdom, they were free to work within other ethical frameworks, especially if the benefits were mutual. It is one of the main sources of disagreement in Bioland; many citizens are unhappy with this and see it as exploitative, while others argue this is condescending to anarcho-evolutionists, who are in no way less capable of developing ethics, but simply have a different system and an ability to tolerate increased human suffering rather than cause further harm to the planet. In the end, anarcho-evolutionists continue to successfully protest their right to experiment, and highlight the losses they would suffer if bioliberals stopped providing relatively advanced and safe facilities for their experimentation. But things can and do go horribly wrong. To prevent this from demoralizing or demotivating its population, the results of failed experiments are celebrated in an annual public Festival of Experimentation and Future Selfhood.

For our part, we just find the zoo lab environment where much of this happens depressing. A grim, concrete enclosure consisting of labs and operating rooms as well as catwalks and alcoves for displaying and inspecting the results. But some of the creatures are delightful and exude a surreal beauty. It's difficult to see it from only one perspective.

DAY 4, RETURN TO DIGILAND AND A VISIT TO THE GOV.PRODUCTS LABORATORY

Digitarians have a government bureau dedicated to noncommercial products designed to meet social and emotional needs deemed of such importance that commercial producers and the market cannot be relied on to meet them. These are usually related to behavioral modification or social conditioning essential to a healthy existence within their society. We were curious to visit one of these labs and discover more. Being a micro kingdom that fully embraces the marketplace as a panacea, these labs are not something they liked to discuss, so we were quite fortunate to be able to visit one.

Approaching in a digicar, we catch our first glimpse of the Gov.Product Laboratory, visible from nearly a kilometer away. Like many digitarian buildings—what would have been thought of as large out-of-town boxes in the old United Kingdom—it sits like an island on a vast plane of tarmac. Having seen many buildings like this while in Digiland, we're excited to be able to enter one at last. As we get closer, we realize it is even larger than we thought, several kilometers long, and very low.

In Digiland, roads, if you can still call them that, are managed in a similar way to the electromagnetic spectrum and telecommunications. "Roads," or more accurately routes, are owned by the government, which leases them to the highest bidder in auctions; once secured, a company then devises tariffs that monetize every square meter and every millisecond they are accessed. Due to the planar nature of the digitarian landscape, several "roads" or pathways can run in parallel, each offering different standards of algorithmic management; some digicars even embody ethiculators, which allow users to select accident resolution algorithms aligned with a specific set of beliefs and ethics that determine who will survive an accident–pedestrian or driver, school bus or goods vehicle.[2]

Once inside the Gov.Product building, we are confronted with a stunning interior landscape. Poured resin floors, fluorescent lighting, and fluoro-colored markings. Slightly raised ramps pass between large concrete cubes, cylinders, and other geometric volumes housing different divisions. We have been told we're heading to the adaptive object division where devices for behavior alignment are developed; we might even get to try a prototype.

Unlike other micro kingdoms, the digitarians still turn to technology as the solution for their society's problems. Political and social forms of solutionism are just too complicated. Technological solutions always feel more promising and are measurable. If the micro kingdom's steering committee frames issues as technological rather than political, they sound more appealing to consumer-citizens and underlying power relations can usually remain intact.

PUBLIVOICE

Our digicar stopped next to one of the largest volumes. Its interior was a cross between a satellite assembly room where strange craft takes place and a minimalist artist's studio space lined with charcoal-colored foam pyramids.

Pacing around the space was a lady wearing a largish cubic volume that rested on her shoulders and enclosed her head. It was made from a pink, translucent material; we couldn't tell if it was fabric or something more rigid, or even a field of some kind. It stayed consistently in isometric projection no matter how she moved. We noticed secondary volumes by her mouth and one of her ears. On its surface, markings subtly changed from mid to dark tones as she spoke.

She was repeating phrases, listening before speaking. Maybe learning them? The phrases didn't quite make sense: "inventory leakage," "rightsizing," "misspeak," "categorical inaccuracy," "here and nowist," "aristorithm," "heritage loan," "bioshuffle," "ethiculator," and "workplay."

Our host told us that the woman was calibrating a publivoice device. Since the recent amendments to the Offenses against the Person Act of 1861, it had become essential for consumer-citizens to familiarize themselves with publispeak, a language constructed almost entirely from euphemisms. It was originally designed for social media, but was now a requirement in all public spaces. If these tests go well, he tells us, then three thousand devices will be produced by Gov.Products and distributed to schools throughout the micro kingdom.

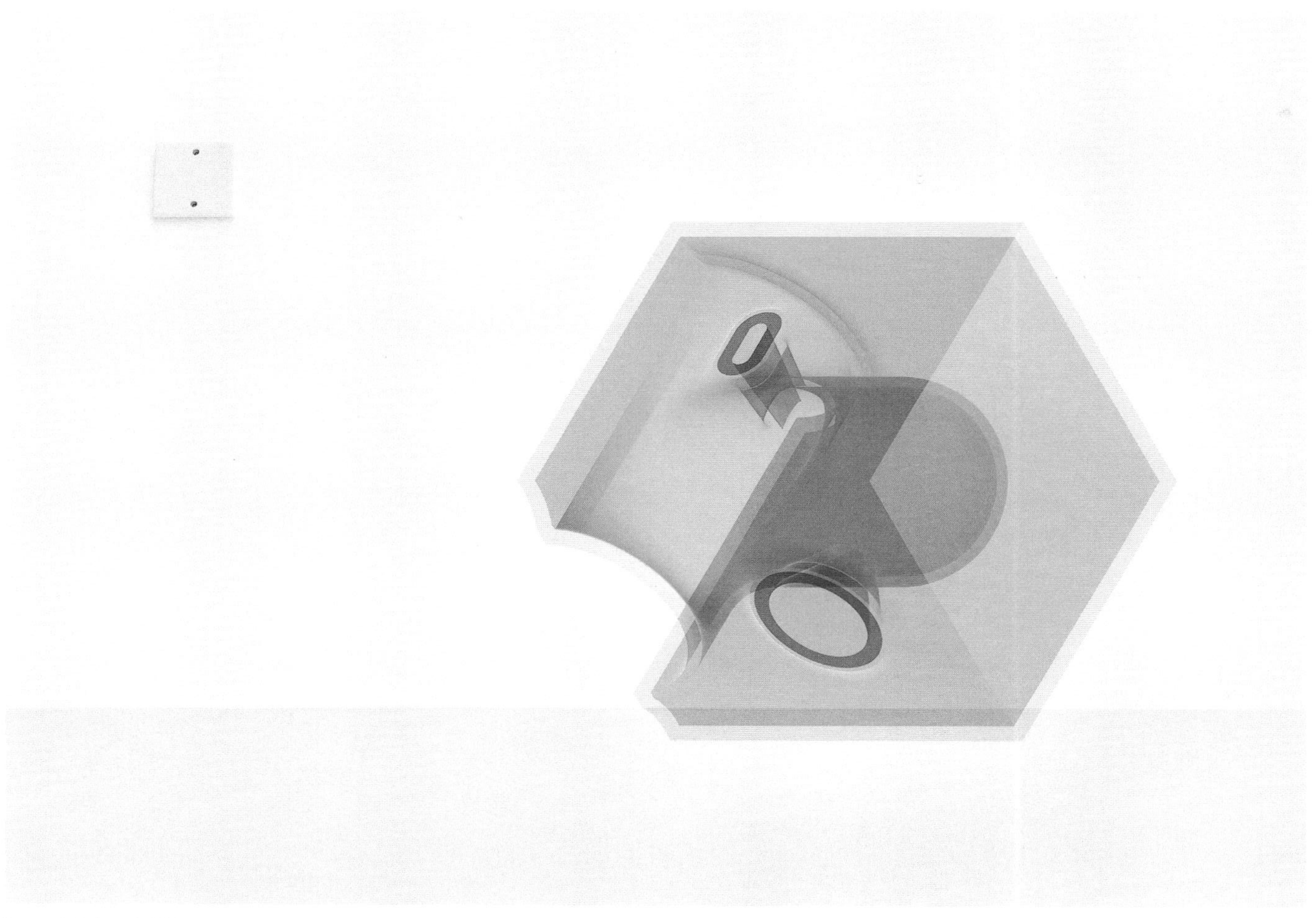

Dunne & Raby, *Publivoice*, 2014, from the series *Not Here, Not Now*. Computer-generated image: Lukas Franciszkiewicz

Dunne & Raby, *Publivoice*, 2015, part of the series *Not Here, Not Now*. Still from video by Keiichi Matsuda. Computer-generated image: Lukas Franciszkiewicz.

In Digiland, lives have been ruined, careers ended, and reputations destroyed by a single decontextualized utterance going viral and crossing the threshold of a thousand "hurts," which obligates regulators to take action and prosecute.

"Too many words are now capable of causing severe emotional harm," our host told us. "It's the act that states, 'Whosoever shall unlawfully and maliciously emotionally wound or inflict any grievous emotional harm upon any other person, shall be guilty of a misdemeanor, and being convicted thereof shall be liable, at the discretion of the court, to be kept in penal servitude for the term of three years, or to be imprisoned for any term not exceeding two years, with or without hard labor.'"

We had so many questions. What about in the home, in the pub, at work, or among friends?

It turns out families are OK as long as they are in an enclosed space that can be regarded legally as the "home" or an extension of it, such as a car, holiday home, or hotel room. It's not OK if visiting another home unless it belongs to category one relatives such as parents, siblings, or children.

At work it's advisable to use publispeak when addressing a group of more than five colleagues, especially if there is an unfamiliar face in attendance whose personal Discomf Index you may not know. But interestingly, there are increasing numbers of underground "speakfreelies" springing up where people go to enjoy emotionally harmful words and expressions.

Just as we were leaving, we saw a list of book titles on her desk familiar to us from our own childhoods—dozens of them. She noticed and asked if we knew what the list was, which we didn't. She explained that nearly all the stories in these books contained words, phrases, and concepts with a Discomf Index in excess of seventy-five, which meant they were likely to be a source of severe emotional distress for many people. "Probably not quite what George Orwell had in mind when he conceived of Newspeak," we heard her say under her breath as we were leaving.

We could have easily walked to the next room, where we were going to see an ethiculator, but we were ushered into a digicar and driven a few meters. Ethiculators are essentially calculators for working out complex (and not so complex) ethical dilemmas. The algorithms are carefully designed. Ethics has essentially become a series of calculations rather than judgments.

In an earlier conversation, we were told that few people in Digiland can imagine instinctively knowing right from wrong. This is due to the UMK's coexisting but very different value systems. Morality is viewed as more fluid and relative, malleable even, so children are taught how to use ethiculators. These evolved from those old computing theories about "moral machines," artificial intelligence, and machine learning. Well, they didn't turn out the way they expected,but they are here to stay now. Without technological assistance, it has become almost impossible to avoid upsetting another person and therefore receiving a heavy fine.

Dunne & Raby, *Ethiculator*, 2014, from the series *Not Here, Not Now*. Computer-generated image: Lukas Franciszkiewicz.

Ethiculators are just the tip of the iceberg. Digitarians have long since given up trying to use judgment; it is just too controversial. No one takes responsibility anymore; if anything were to go wrong, they might end up being crushed by the system. Trust seems like a quaint concept from a bygone era. Historically, they had guidelines, then templates, then algorithms, and then finally ethiculators. People said they were much happier. They just accepted what the device told them. It is, after all, a product of the most advanced programmers, philosophers, mathematicians, and ethicists, all highly regarded disciplines and professions in Digiland.

Although both these products are specific to Digiland, some of their inventions could work elsewhere. Digitarians are highly sensitive to the practical rather than privacy implications of permanent digital traces and have developed a technology that allows a digital text's owner to retrospectively erase it. Of course there are rules, legal obligations, and countermeasures to control its misuse, but essentially it consists of a typeface with an inbuilt kill switch. For example, one's textual traces could be programmed to be wiped clean on a certain date or its erasure could be triggered by an event such as the closing of a company. Once published, it is illegal to change the kill date of a font without permission. Some typefaces have a lock for use in contracts and other legal documents.

DAY 5, RELAXING WITH ANARCHO-EVOLUTIONISTS

We were invited back by one of the anarcho-evolutionists we met at the zoo lab to accompany a group of cyclists on their Very Large Bike (VLB). The family or clan is the most important anarcho-evolutionist unit. Families evolve around particular forms of transport using a combination of genetic modification, training, and the passing down of knowledge and skills from generation to generation. A distinctive physique is associated with each clan and is a matter of pride. Cyclists have well-developed thighs, while balloonists are tall and willowy. Our new contact is a second-generation cyclist, with abnormally bulky thighs.

Dunne & Raby, *Anarcho-Evolutionist Sport*, 2014, part of the series *UMK: Lives and Landscapes*. Illustration: Miguel Angel Valdivia.

There have been quite a few studies made into the rituals of the anarcho-evolutionists, but one of our favorites is on "greetings" by an anthropologist.[3] She observed that greetings are an opportunity to measure evolutionary progress. When members of the same clan greet one another, they compare the physical attributes that they have evolved to accommodate their method of transportation: for the balloonists, a tall, lean frame, and for the cyclists, well-developed thighs. Those who are evolutionarily "older" have inherited the most adapted features. Balloonists compare the size of their feet by placing them side by side. Men touch the outside of their feet together, women the inside, and men and women simply touch toes, as sexual dimorphism removes the need for direct comparison. Foot size, as an indicator of height, reveals the evolutionary age of the individual. Cyclists compare the strength of their thighs by lunging side by side. Like the balloonists, men touch the outside of the lunging leg, women lunge touching the inside of the leg, and men and women touch knees. Those with the deepest, most sustained lunge are the "oldest."

Anarcho-evolutionists have also developed several animal variations to satisfy their practical and emotional needs. They spend a great deal of time with their animals and need to feel some kind of bond; they are almost team members. The hox is a mix of horse and ox, a hybrid animal bred to move heavy loads and pull carriages, while the pitsky is a combination of pit bull terrier and husky, designed for pulling smaller loads and personal protection.

The anarcho-evolutionist's world is a world without cars. Their transport is either human, wind, or (genetically modified) animal powered. The vehicles are designed around the principle of organization without hierarchy, and embody their social order and values. Sociability and cooperation are more important than speed and competitiveness. They travel in groups, each doing what they are best at, and each is responsible for a part of the vehicle. The VLB is designed for traveling long distances in groups of twenty or so, pooling efforts and resources. Traveling on abandoned motorways, it is gently steered by leaning, with each person knowing from experience and practice just how much is required of them. While the elderly, young, and less able are carried along by the others, their role is to sing and tell stories, providing entertainment and motivation for the rest of the group. Face-to-face conversation and flat hierarchies are important, and the layout of the VLB reflects this. Animals are often passengers too.

Having spent several hours on one, we can confirm that traveling by VLB is a wonderful experience. One of the most memorable aspects of the journey was watching anarcho-evolutionists interact. They use exaggerated facial expressions and an extended range of sounds. And they thoroughly enjoy rhyming and onomatopoeia. Inspired by animal sounds, they sometimes modify their vocal cords. Warning sounds for their vehicles are also made by people, mainly out of sheer joy and exuberance. There is a constant acknowledgment of others through micro sounds, almost like acoustic winks and nods. An incredible range of sound is produced—super fast as well as slow, or stretched, used to speak of qualities that our language rarely captures.

This fascination with sound and collective activities extends to their musical instruments and entertainment. For example, when we were traveling with the cyclists, they met up with another clan that had built a fascinating musical instrument that required seven players. At one end were blowers with well-developed lungs, and further along were “flute fingerers” with seven or more fingers on each hand.

We were also able to watch one of their sporting events. Hundreds of men and women running about in a field during a storm. Each had what looked like a parachute strapped to their back, and several had been lifted off the ground and looked as though they were about to be swept away, but others were pulling at fine cables, ensuring they remained tethered. Like most national sports, it says much about what they value.

DAY 6, THE COMMUNO-NUCLEARIST TRAIN

Once back in Digiland, our next destination was the communo-nuclearist train.

We were looking forward to meeting the communo-nuclearists so much. It's been difficult to arrange; being self-contained, they do not really interact much with the other micro kingdoms beyond the quinquennial re-signing of the UMK Memorandum of Tolerance and occasional trading.

Luckily, our host had an old communo-nuclearist map-calendar that shows exactly where they will be at any point in the year. Their calendar is a thing of beauty. Time has been reclaimed and reorganized to suit their mobile, circuitous existence. Their year consists of 360 days, made up of 12 months, each with 30 days. One week is 10 days, or the time it takes to make one 960-mile rotation traveling at 4 miles per hour. Like traditional Japanese calendars with 72 micro seasons, their seasons are many. Early summer, late summer, early autumn, late autumn, early winter, late winter, early spring, and late spring.

While waiting for the train at our rendezvous, we asked our host about digitarian names, which are made from the "spelling alphabet," or the "NATO phonetic alphabet" as it is more commonly known. Our host, for example, is simply called Charlie-Echo-Delta, his partner is Golf-Bravo-Kilo, and their child is Alfa-Juliett-Foxtrot. They are wonderful names, especially when spoken in the lyrical eastern digitarian topolect found in Digiland Zone B. He tells us that digitarian names evolved from the amount of time they spent communicating with call centers in the early days. It began with sequences of letters that

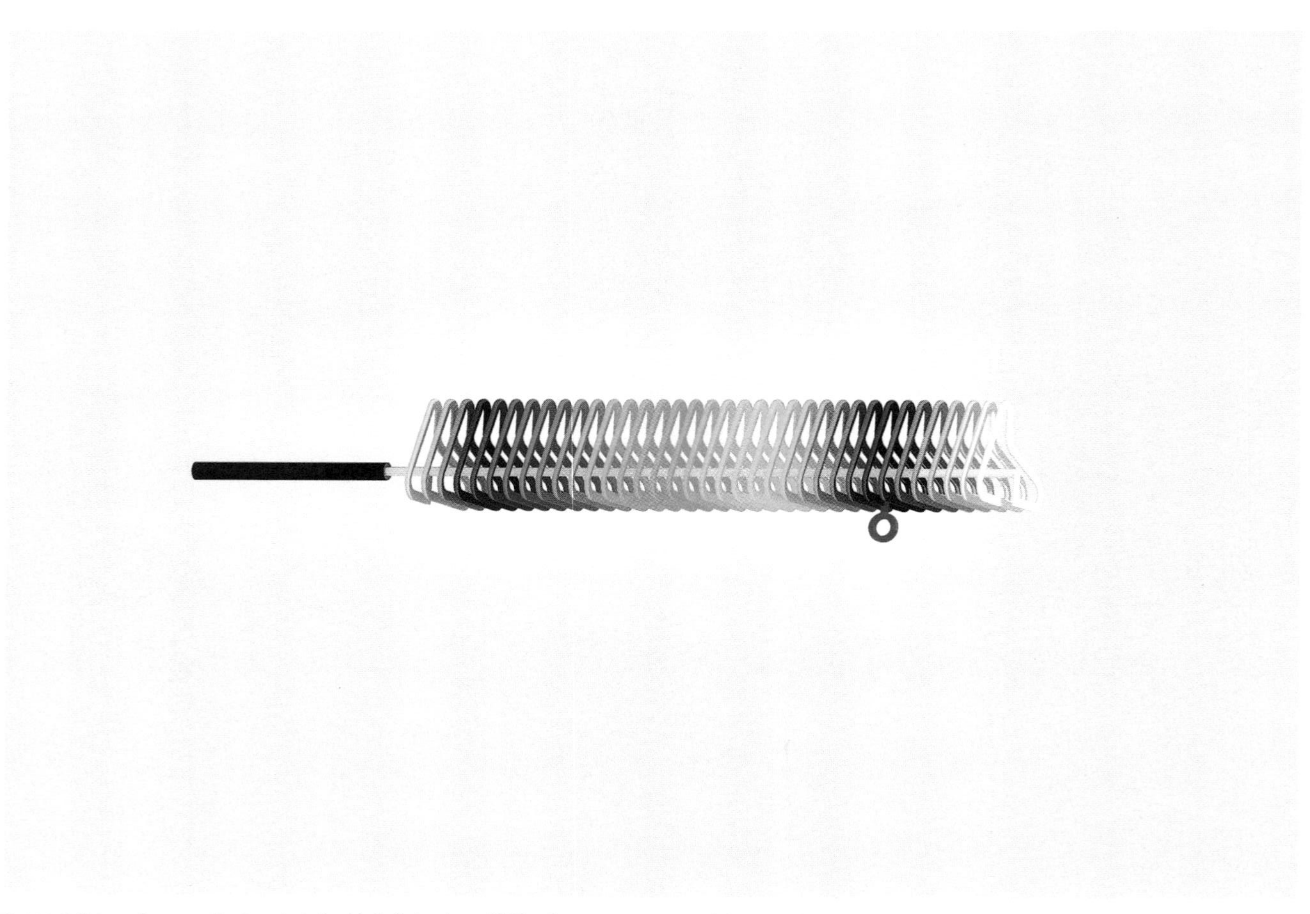

Dunne & Raby, *Communo-Nuclearist Spatial Calendar*, 2023. Computer-generated image: Franco Chen.

spelled recognizable names (to us), but over time they gradually began to be arranged in sequences that produced the most pleasant or distinctive sounds and rhythms.

Suddenly the train appeared. It was magnificent, comprised of 75 carriages, each measuring 130 feet (40 meters) in length by 65 feet (20 meters) in width. Straddling two sets of 10-feet-wide tracks, it travels at 4 miles per hour (6.4 kilometers per hour) and never stops. Meetings with communo-nuclearists tend to happen next to the tracks, within 30-minute segments, or the time it takes for the train to pass a point from front to back. Inhabitants live inside the mountain carriages, which also contain labs, factories, hydroponic gardens, gyms, dorms, kitchens, nightclubs, and everything else they need. On its mountains are swimming pools, fish farms, and bookable huts for periods of isolation. The environment surrounding the tracks, like a demilitarized zone, has become a natural paradise, a wilderness to be enjoyed by nature-loving communo-nuclearists from the safety of their train. Catching a glimpse of them as they pass is quite something. Their clothing is derived mainly from indoor styles such as pajamas, slippers, and dressing gowns, but made from outdoor materials such as dyneema, a composite fabric. They kindly gifted us a set, which we will donate to our local lending library of things when we return home.

On the train, no one speaks about the future, just as no one speaks about growth. Somehow they live in a perpetual present. This is partly due to its fixed population, but also because of its fixed route. When you ask a trainlander what they think about the future or what their hopes for the future are, they typically say, “Like now.”

Communo-nuclearists have a refined sensitivity to different shades of reality: possible and impossible, imagined, actual, virtual, and so on. There are many occupations related to this under the general title of reality constructor, which includes reality designers, reality producers, new reality finders, and reality fabricators (bottom-up and top-down). These highly valued occupations help the community transcend the limits of train reality. They are not exactly escapists, but they do push fiction to extremes. The highest ranking of these occupations is a connoisseur of unreality, a role dedicated to the materialization of truly impossible objects, like those found in Borge's "Tlön, Uqbar, Orbis Tertius," made from a combination of sounds and visual qualities, or impossible colors lying beyond human perception, or even extreme mathematics and phenomena related to quantum mechanics. To communo-nuclearists, a description of an object is equal to the thing itself, and devices like perpetual motion machines have the same importance as conventional machines. Their reasoning is, if an object is capable of being imagined, then it exists, at least in the mind, which is part of the world. Much of their time goes into imagining impossible objects and developing strategies for their materialization while dreaming of adding new subcategories to Meinong's theory of objects. Most fail, of course, but it is the attempt that is valued most of all.

This fascination with nonexistent, or more accurately, imaginary objects, probably has something to do with the constraints their limited space places on the number of physical objects that can exist in the train at any one time. If physical space becomes available, then a new object can be imagined, made, and stored in their famous Lending Library of Things. On the train, everything is shared and used as needed.

Living within a confined physical environment has led to another fascination, the faraway, and a culture of observation and classification. Over time, they have become experts in the production of optic devices known by digitarians as peepitubes: telescopes, binoculars, camera lenses, and microscopes. It is the basis of their trade with the digitarians, especially for school curricula and the study of color.

Dunne & Raby with Sarah Hawes, *Communo-Nuclearist Pajamas (Trousers)*, 2023. Photograph: Yuhan Pan with Yoshe Li and Billy Chen.

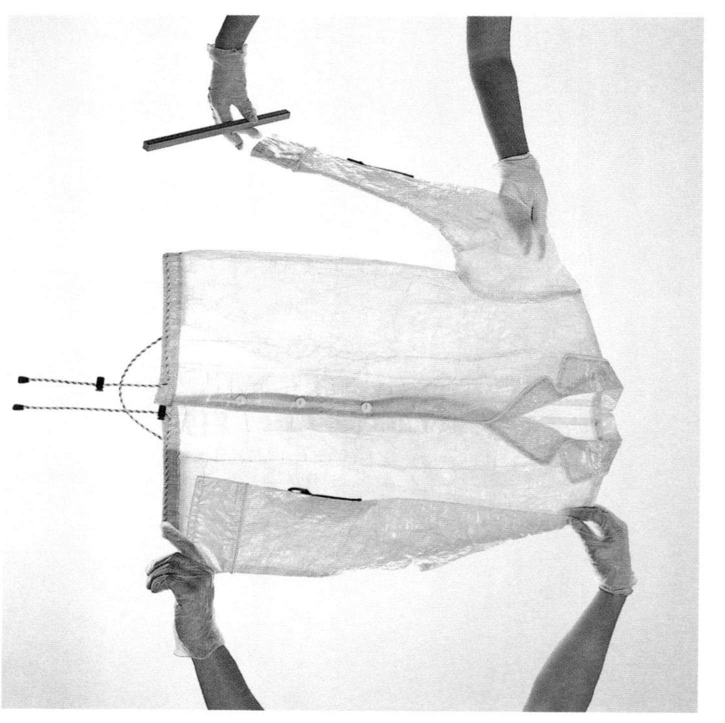

Dunne & Raby with Sarah Hawes, *Communo-Nuclearist Pajamas (Jacket)*, 2023. Photograph: Yuhan Pan with Yoshe Li and Billy Chen.

Dunne & Raby with Sarah Hawes, *Communo-Nuclearist Pajamas (Hood)*, 2023.
Photo: Yuhan Pan with Yoshe Li and Billy Chen.

Dunne & Raby with Sarah Hawes, *Communo-Nuclearist Pajamas*, 2023.
Photo: Dunne & Raby.

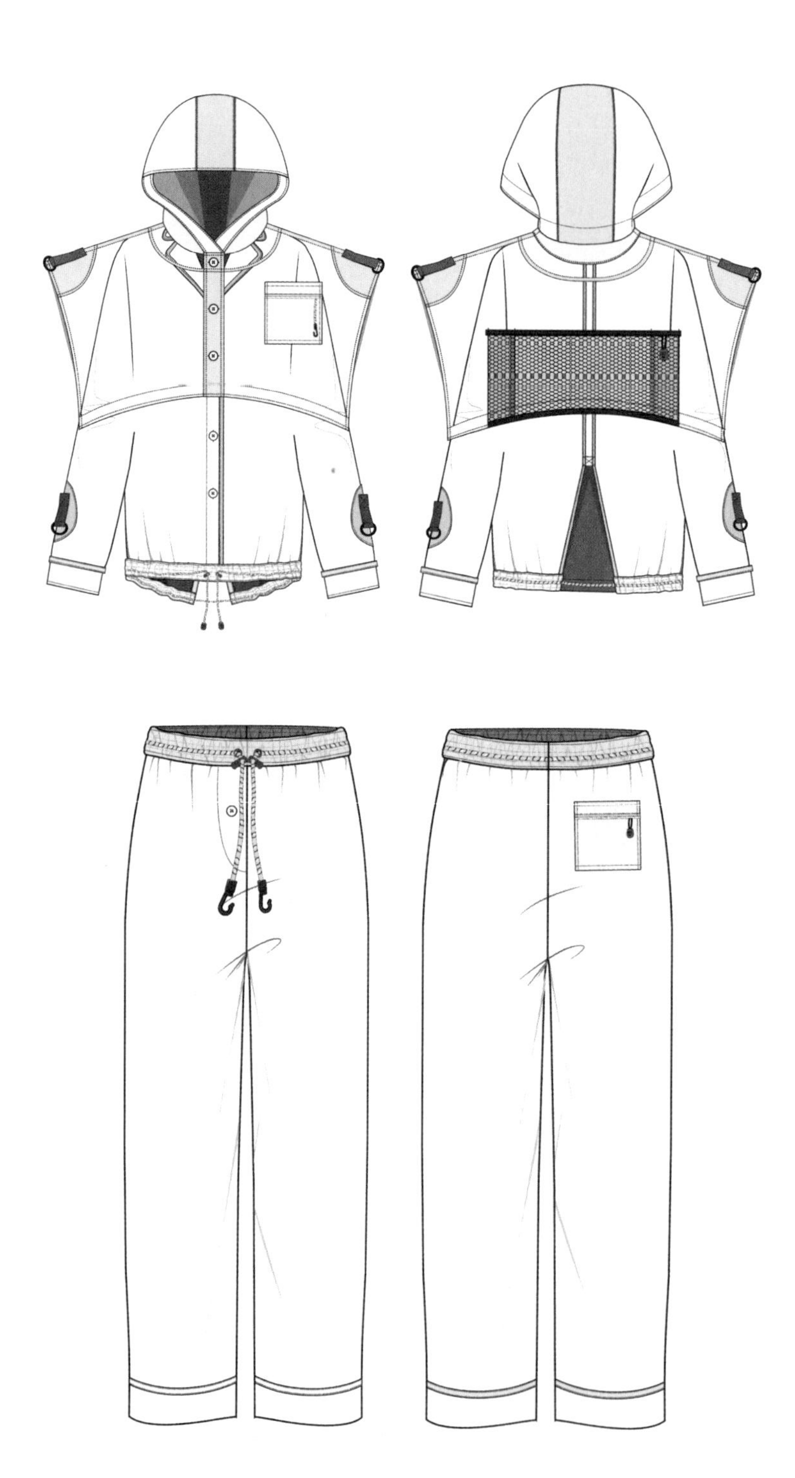

Dunne & Raby with Sarah Hawes, *Communo-Nuclearist Pajamas*, 2023. Illustration: Sarah Hawes.

DAY 7, ESCALATOR MOUNTAIN

While visiting the communo-nuclearists, our guide received a message about an important event in Digiland: the unveiling of a design for a major public project that symbolized core digitarian values. We had to leave immediately if we were to see it.

Dunne & Raby, *Escalator Mountain*, 2018/2024.

The presentation was at the Bureau for the Production of Cultural Metonyms. It turned out they were presenting a massive scale model of Escalator Mountain . Once the model was unveiled, the excitement in the room was palpable. Although we admired the ambition and technical prowess of building a mountain from scratch, something about it made us uncomfortable. We think it was the scaffold of escalators encasing the mountain that was most disturbing. The tourist assumed a passive role, enjoying the views the state had designed for them. In many ways it is the opposite of a roller coaster–calm, safe, and a little bit boring, but in a land with no construction higher than ten meters, that is exactly what they want. Simply being raised above the tarmac plains is a novelty, especially in a micro kingdom obsessed with health and safety. When you compare it to the efforts of our own governments to promote their "vision" through extravagant public projects, it sort of makes sense, but still...

7 Once Po
Now Impos
A Partial
Inventory
National
Made Phys

ssible,
sible:

of
Dreams
ical

Chapter 7

Once Possible, Now Impossible: A Partial Inventory of National Dreams Made Physical

We ended *The United Micro Kingdoms, a Travelers' Tale*, with a visit to an event celebrating a new megaproject called *Escalator Mountain*. In keeping with the values underlying life in Digiland, the mountain was a celebration of passivity and control. Its surface was clad in a scaffold of escalators taking visitors to its summit, from which they could take an elevator back to the ground level. As with many of these megaprojects, it is a product of a specific political ideology—in this case, a kind of digitally mediated transparent and participatory democracy built around surveillance. Many projects on this scale are created by governments with mildly authoritarian leanings, or technocracies, but direct democracy has its issues too. Nicely illustrated by a UK government agency's approach to choosing a name for a new $287 million polar research ship that used a web-based polling campaign called #NameOurShip where the public could suggest and vote for names. The winner was RRS *Boaty McBoatface*, which despite being the most popular choice, was unsurprisingly rejected. RRS Sir *David Attenborough* was chosen instead, but *Boaty McBoatface* did end up being used for one of its submersibles.

Like many symbolic megaprojects, *Escalator Mountain* seems pointless, vain, and extravagant; a monument to missed opportunities. In this case, that is exactly what it is meant to be: an absurdist embodiment of so much that is wrong with many of these kinds of megaprojects. While a mountain might seem a little extreme, even for a symbolic project, we were surprised to learn there have been quite a few proposals for artificial mountains. There is the Berg at Tempelhof in Berlin, a 1,000-meter-high leisure and natural zone that took on a second life in Elvia Wilk's novel *Oval*.[1] Or the United Arab Emirates' proposal for a mountain to increase desert rainfall. And Die Berg Komt Er, a 2-kilometer-high, 5-kilometer-wide leisure mountain proposed by a journalist in the Netherlands.[2] All of these underwent feasibility studies, and in the latter case, was also viewed as an intellectual challenge and platform for innovation. Besides these, there are human-made mountains produced by waste from large-scale industrial processes, such as the cement and steel industries. In Loos-en-Gohelle, an ex-mining area in France, mine 11-19 has two "terrils," which reach a height 146 meters, forming the largest slag heaps in Europe.[3] It is now on the UNESCO world heritage list, and provides a home to many different plants and wildlife.

But our favorite, for its audacity alone, has to be a proposal from the 1970-1980s by Australian Lawrence J. Hogan called the Engineering Australian Plan that included a 1,780-kilometer-long human-made mountain range, situated at a longitude of 130 degrees, 4,000 meters high, with a 2,000 meters-high plateau and 10,000-meters-deep base. On learning about our comparatively modest proposal, our New School colleague McKenzie Wark gifted us Hogan's wonderful book *Man-Made Mountain* about the background, details, and hopes for the project.[4]

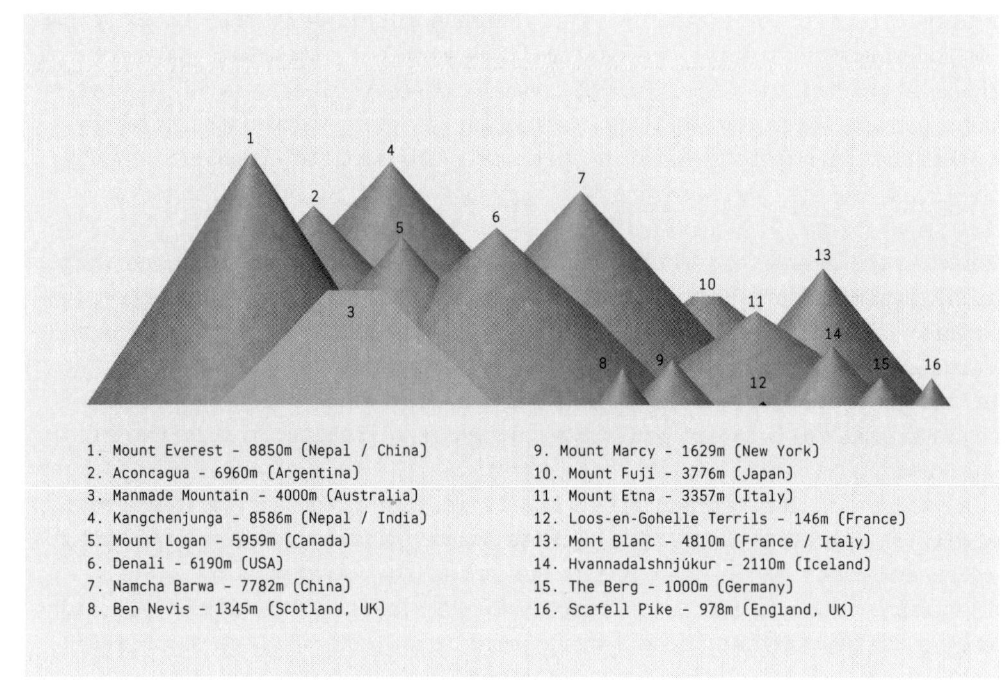

Adapted from Lawrence J. Hogan, *Man-Made Mountain* (Sydney: Charter Books, 1979).

The mountain part was intended to benefit "climate, industry and environment," specifically by upgrading precipitation, "which will rinse out moisture from the air mass which crosses the Great Dividing Range." But it had many other purposes central to the Engineering Australian Plan: providing storage space for national product inside the mountain; offering shelter for people if ever necessary; hiding industries from sight within the depths of the mountain; providing rain by raising clouds and transforming an area into a sort of garden of eden; and supplying work for millions of people in every trade and profession.

Although so far, it has not been possible to build an artificial mountain beyond the scale of what might more accurately be described as an artificial hill or mound, human-made mountains dramatically capture how megaprojects, through their scale, use of resources, and physicality, impact the world, reshaping reality, not just in the present, but over extended timelines, rendering many other future possibilities impossible. In an interview about his book *On the Future*, Martin Rees, astronomer and cofounder of the Centre for the Study of Existential Risk at Cambridge University, remarked that although things are better for many people today than they were hundreds of years ago, the gap between how things are and how they could be is far greater today than it has ever been.[5] It's an interesting thought. How can that be? Does it mean that since medieval times, there are now more possible worlds available to us or our imaginations? That today, we somehow live within a conceptual multiverse where if we had the will, things could be radically otherwise? In Rees's view, one of those worlds is one where part of the extreme wealth of the world's thousand-richest people could be used to dramatically transform the lives of the "bottom billion."

This section is not a celebration of megaprojects but instead uses them to highlight forms of impossibility that shift in and out of focus with the passage of time. In *Extinct: A Compendium of Obsolete Objects*, the authors write about how objects that have slipped out of existence speak of worlds that once were, but are no more. At the scale of very large objects, the objects still exist, but it is even more clear that the worlds in which they belong no longer exist. Located at the juncture of world making, and the production and destruction of realities, they are monuments to lost or obsolete realities as well as extinct worlds. They continue to physically exist—unlike products that are taken out of circulation for a variety of reasons—act on, and distort whatever new reality prevails. They also act on the imagination to influence what is possible, and more usually, what is not.

Megaprojects, or very large objects, as we like to call them, can take many forms: the reunification of Germany, Brexit, the Catholic Church, China's Belt and Braces initiative, large-scale urban developments such as King Abdullah Economic City or Songdo in South Korea, technological objects such as the International Space Station, and ancient examples such as the pyramids, Great Wall of China, or Stonehenge that even today are not fully understood. There is a fine line between projects where the objects are incidental to a system or by-products of it, and objects that require a system to enable them or for them to work. Of course there will always be a gray area, but our focus is on objects that stand out in relation to their support systems. Even if the "object" is an intangible idea, like a nation. All of them represent dreams of some kind, made real. These very large objects are not quite what Timothy Morton has described as hyperobjects either, but they are relatives. Hyperobjects are too vast and too complex to fully grasp in our human-sized imaginations, and often lack clarity on where they end and something else begins. Climate, the internet, and plastic in the ocean are examples.

Most interesting for us is their status as seemingly impossible objects, at least by today's standards. Most of these very large objects could only be imagined and realized in a specific historical moment. Before and after that moment, they are rendered unreal or maybe unrealistic, and become impossible, whether for ideological, functional, social, technical, or economic reasons. As well as appearing to be seemingly impossible objects from today's perspective, they render many other potential very large objects impossible, forever destined to occupy the imagination. They are like monuments to the ever-shifting contours of possibility. Material anachronisms representing not just other times but also other ways of seeing, making sense of, and especially being in the world, made tangible at a vast scale. A scale that shapes reality from that moment on, closing down other possibilities that might once have been possible, but are now rendered impossible. Concrete expressions of what design theorist Tony Fry has called defuturing. They mark a weird conceptual frontier where reality is reconceptualized and remade in slow motion; where ideas shimmer between crazy and real, between doable and ridiculous, fueled by political will, ego, technological breakthroughs, propaganda, and megalomaniac imaginations.

To us, this suggests a world built from ideas, not just stuff. Each one of these very large objects was only possible to imagine and implement within a specific intellectual framework or episteme. Where certain ideas aligned with prevailing social, political, and economic realities. It's too easy to look back, especially at some recent very large objects from the twentieth century, and believe we are still living in the same reality; that there is conceptual continuity. Many of them marked an ontological shift that permanently affected what could exist afterward. Think of the US highway system, for instance, and how the world it led to makes imagining how things might have been otherwise almost impossible. As historian Quentin Skinner writes,

> [The] historian can help us appreciate how far the values embodied in our present way of life, and our present ways of thinking about those values, reflect a series of choices made at different times between different possible worlds. This awareness can help to liberate us from the grip of any one...account of those values and how they should be interpreted and understood. Equipped with

a broader sense of possibilities, we can stand back from the intellectual commitments we have inherited and ask ourselves in a new spirit of enquiry what we should think of them.[6]

There are many very large objects that could have been included in this inventory. In the end, the ones we chose are extreme in at least one dimension. It might be physical size, ambition, the time taken to realize it, its cost, or the number of people involved; some quality that locates it on the edge of possibility. Although cost is just one factor, once one gains a basic awareness of what can be done for how much and at what scale, or what a billion or trillion dollars means in terms of remodeling reality, the US$44 billion that Elon Musk paid for Twitter suddenly takes on a different meaning when compared to what else might have been.

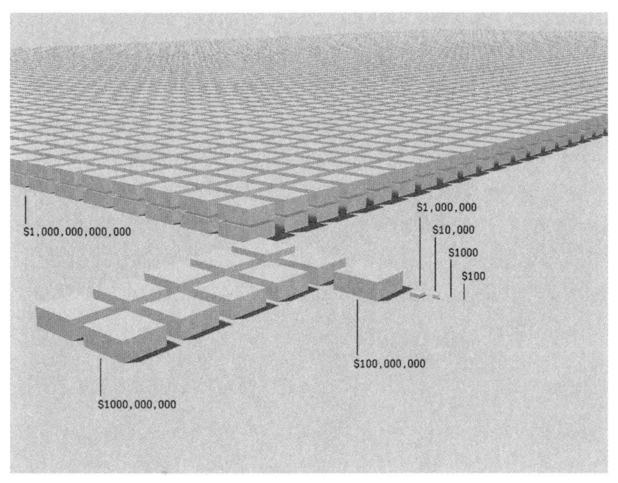

One Trillion, One Billion, One Hundred Million, One Million, Ten Thousand, One Thousand, One Hundred USD. Detail. Adapted from http://www.ewan.dk/1trilliondollars.

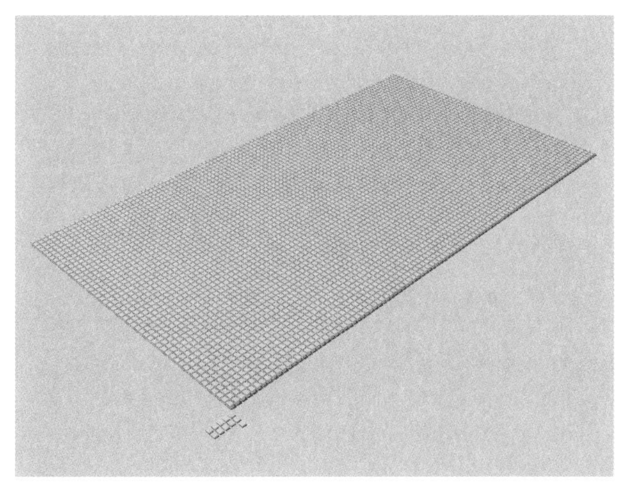

One Trillion, One Billion, One Hundred Million, One Million, Ten Thousand, One Thousand, One Hundred USD. Adapted from http://www.ewan.dk/1trilliondollars.

Today, it feels as though we are living in a time where little is possible. At the exact moment some kind of planetary unity is needed, or at least a shared planetary imagination, a multipolarized fractiverse is emerging, compounded by an atmosphere of wariness toward anything that might be thought of as grandiose. As writer Alec Nevala-Lee mentions in "The Dreamlife of Engineers," climate-related megaprojects continue to circulate in the media, on standby until we no longer have a choice due to a failure of political will and imagination.[7] Ideas range from covering an area the size of Nigeria with solar panels, to planting a trillion trees to control atmospheric carbon levels, to spraying solar aerosols into the atmosphere to reflect sunlight away from the earth. But as Elizabeth Colbert says of carbon removal, "One of the peculiarities of climate discussions is that the strongest argument for any given strategy is usually based on the hopelessness of the alternatives: this approach must work, because clearly the others aren't going to...As a technology of last resort, carbon removal is, almost by its nature, paradoxical. It has become vital without necessarily being viable. It may be impossible to manage and it may also be impossible to manage without."[8]

In light of this, maybe the very large object needed in contemporary times is not physical but instead conceptual, requiring a radical undoing of current mindsets to enable us to collectively set off in new intellectual, creative, and imaginative directions less harmful to the planet itself as well as the multitude of life-forms we share it with. A new worldview or mindset as megaproject.

The Inventory

The following models of various national and international dreams made physical mark the ever-shifting borders between the possible and the impossible over time. Simultaneously ushering new realities or worlds into existence, while foreclosing others.[9]

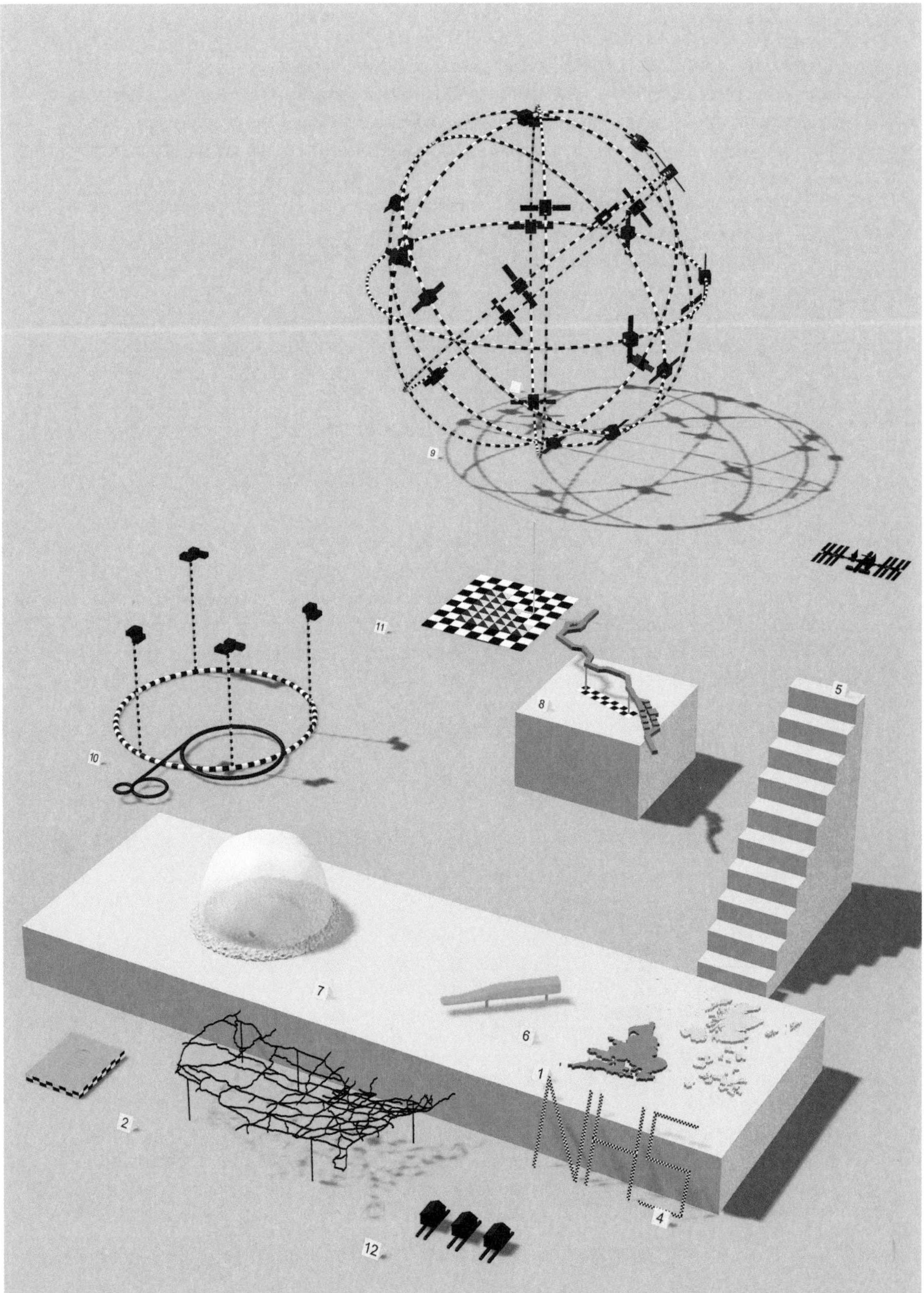

A Partial Inventory of National Dreams Made Physical. Computer model: Franco Chen. Illustration: Dunne & Raby.

Brexit

On Friday, June 24, 2016, sixty-seven million people woke up in one of two United Kingdoms.[10] Each with its own imagined history and future. Overnight, the United Kingdom effectively split into two worlds, proving for half the population that the seemingly impossible is actually possible.

Ideology: constitutional monarchy and parliamentary democracy
Time frame: 2016-2020
Labor force (citizens): 67 million (approx.)
Cost: estimated to be US$124 billion per year (Financial Times, 2022)

UK GDP: $3.071 trillion (2022)[11]
UK GDP per capita: $45,485 (2022)

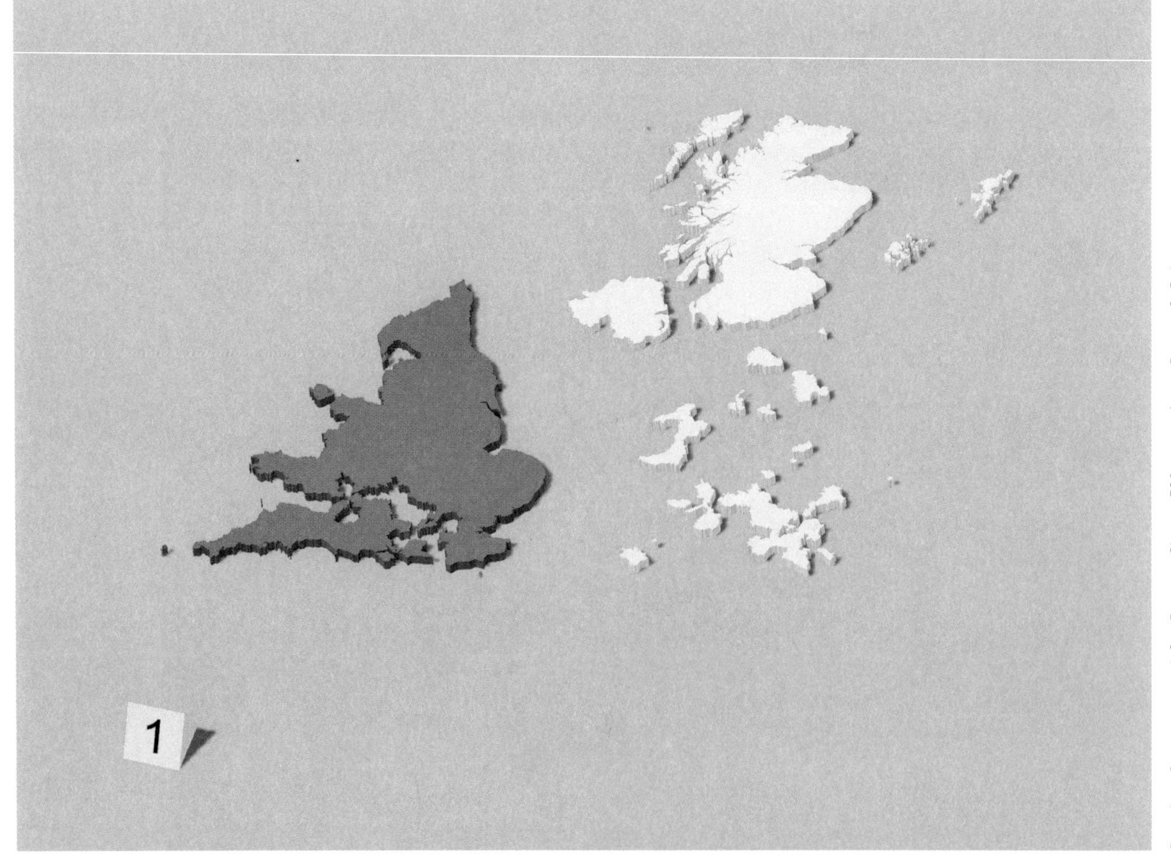

Brexit. Computer model: Franco Chen. Illustration: Dunne & Raby.

US Interstate Highway System

It had a practical aim. To reduce distances between key cities as well as speed up the movement of goods and people across the country. But once in place, its forty-one thousand miles dramatically accelerated the evolution of automobiles and our dependence on them, at least in the United States. Even today, dealing with the hybrid human-car being it produced is problematic, especially in the face of global warming and the role of fossil fuels.

Ideology: Dwight David Eisenhower period of US liberalism
Time frame: 1956-1992
Labor force: not known
Cost: US$114 billion (equivalent to US$597 billion in 2022)

US GDP: $25.463 trillion (2022)
US GDP per capita: $75,269 (2022)

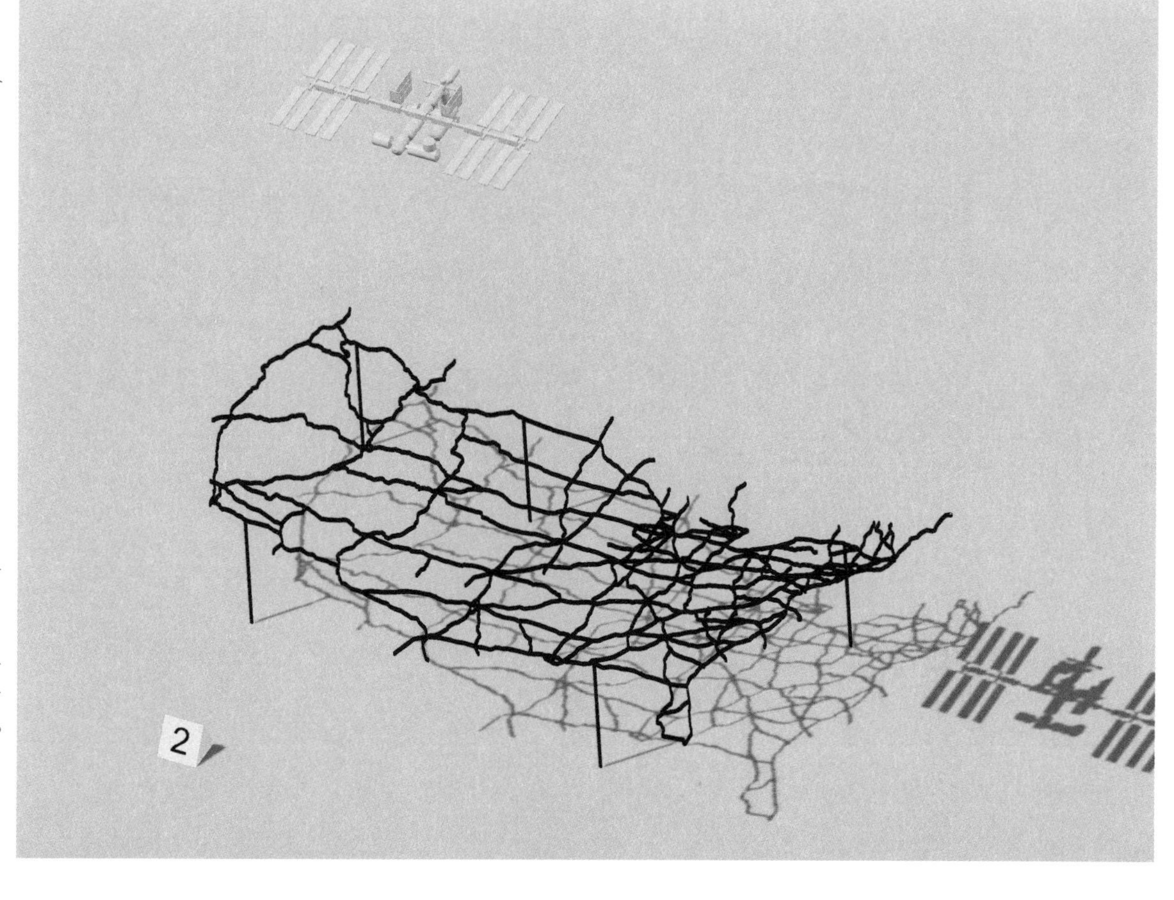

Interstate Highway System. Computer model: Franco Chen. Illustration: Dunne & Raby.

NASA Apollo Program

A simple goal. Place a human being on the moon and take a photograph. But in the process, it also revealed earth itself to be an object, leading to a new collective imaginary.

Ideology: Cold War
Time frame: 1961-1972
Labor force: 300,000-400,000
Cost: US$25.4 billion (1973),
US$283.8 billion (2019)

US GDP: $25.463 trillion (2022)
US GDP per capita: $75,269 (2022)

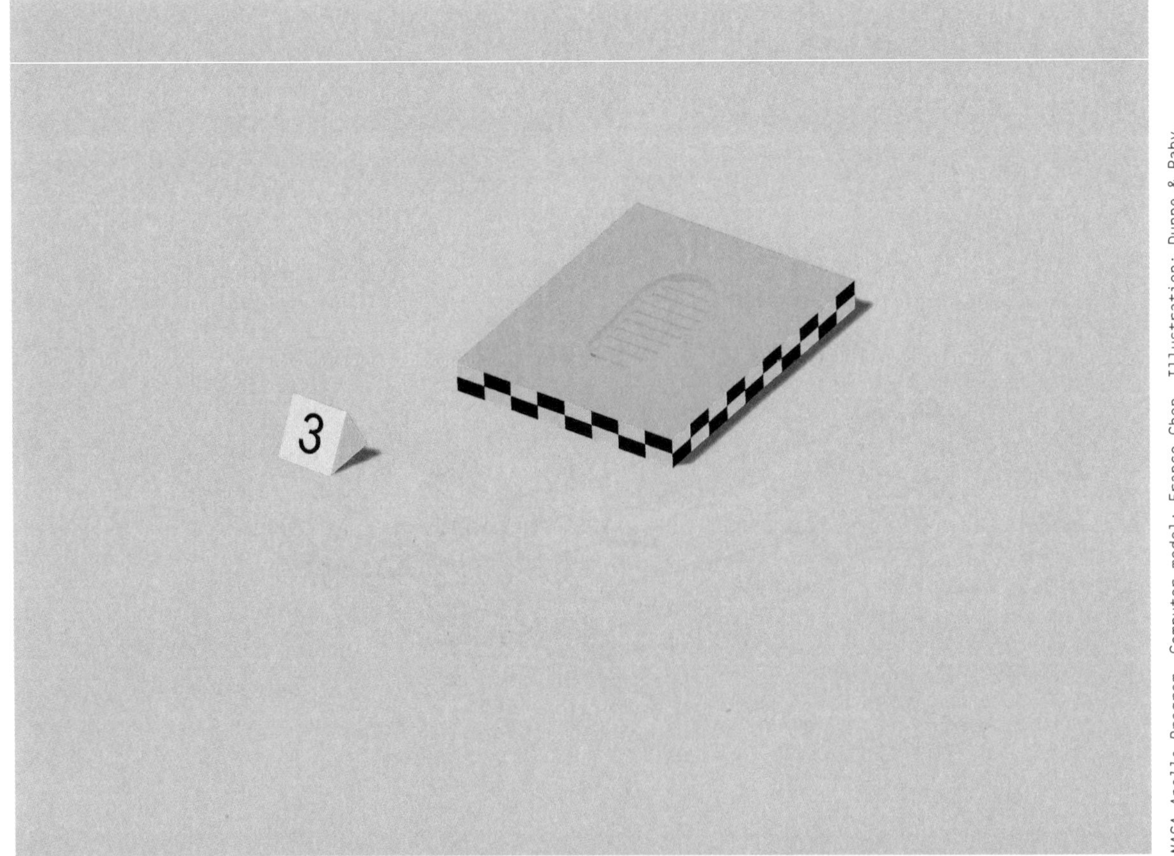

NASA Apollo Program. Computer model: Franco Chen. Illustration: Dunne & Raby.

UK National Health Service

An idealistic postwar dream made real by providing free health care for all in the United Kingdom. Marked a dramatic shift from a Victorian to a modern mindset. Today, it is one of the world's largest employers and a symbol of so much that is important to the United Kingdom. But for how long?.

Ideology: labor government / social democratic welfare state and constitutional monarchy
Time frame: 1948 >
Labor force (employees): 1.4 million people (2023)
Cost (annual): US$180 billion in 2022-2023

UK GDP: $3.071 trillion (2022)
US GDP per capita: $45,485 (2022)

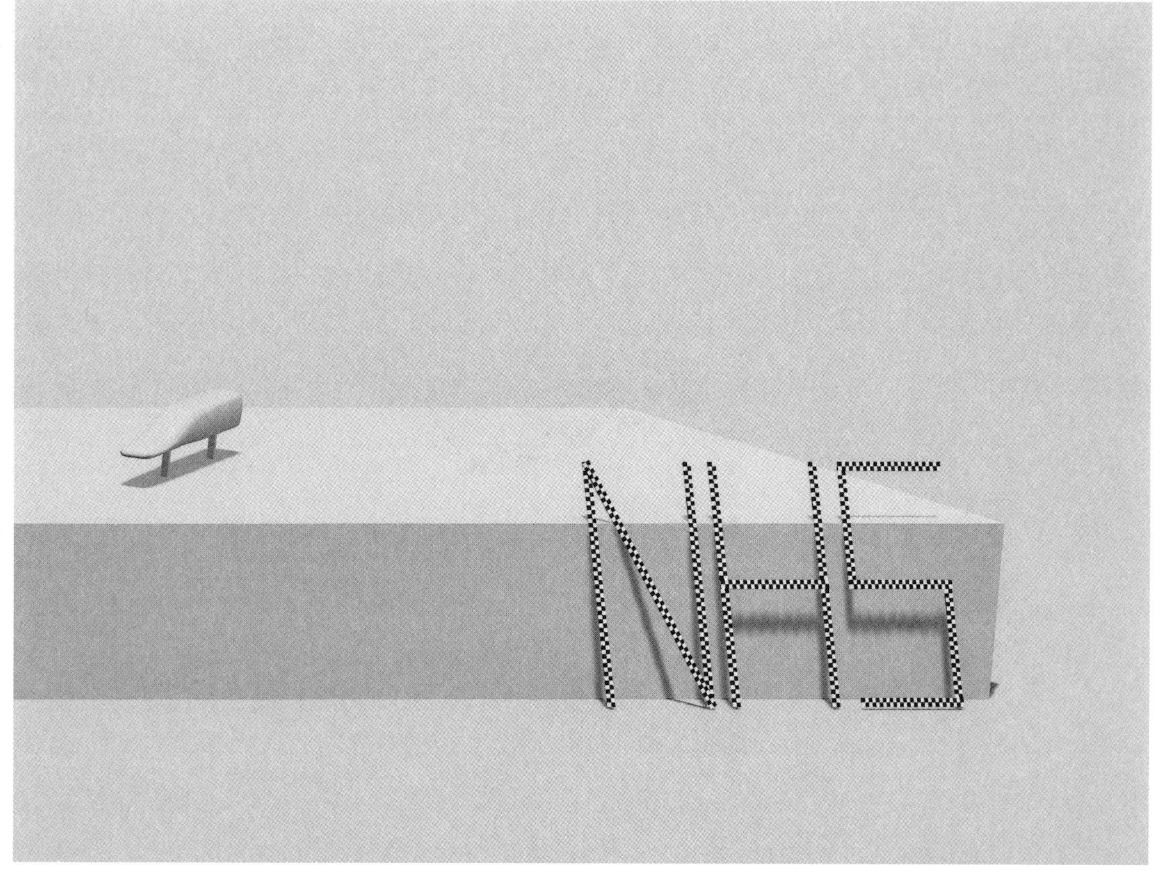

National Health Service (NHS). Computer model: Franco Chen. Illustration: Dunne & Raby.

International Space Station

Possibly the most straightforward very large object in the inventory; it is big, spectacular, and expensive. In fact, it is often cited as the most expensive object ever made.

Ideology: internationalism; cooperative program between Europe, the United States, Russia, Canada, and Japan.

Time frame: Ronald Reagan directive in 1984, assembled 1998-2011
Labor force (visitors): 269 individuals from 21 countries have visited the International Space Station
Cost: US$150 billion (2010) + US$3 billion to operate per year

US GDP: $25.463 trillion (2022)
US GDP per capita: $75,269 (2022)

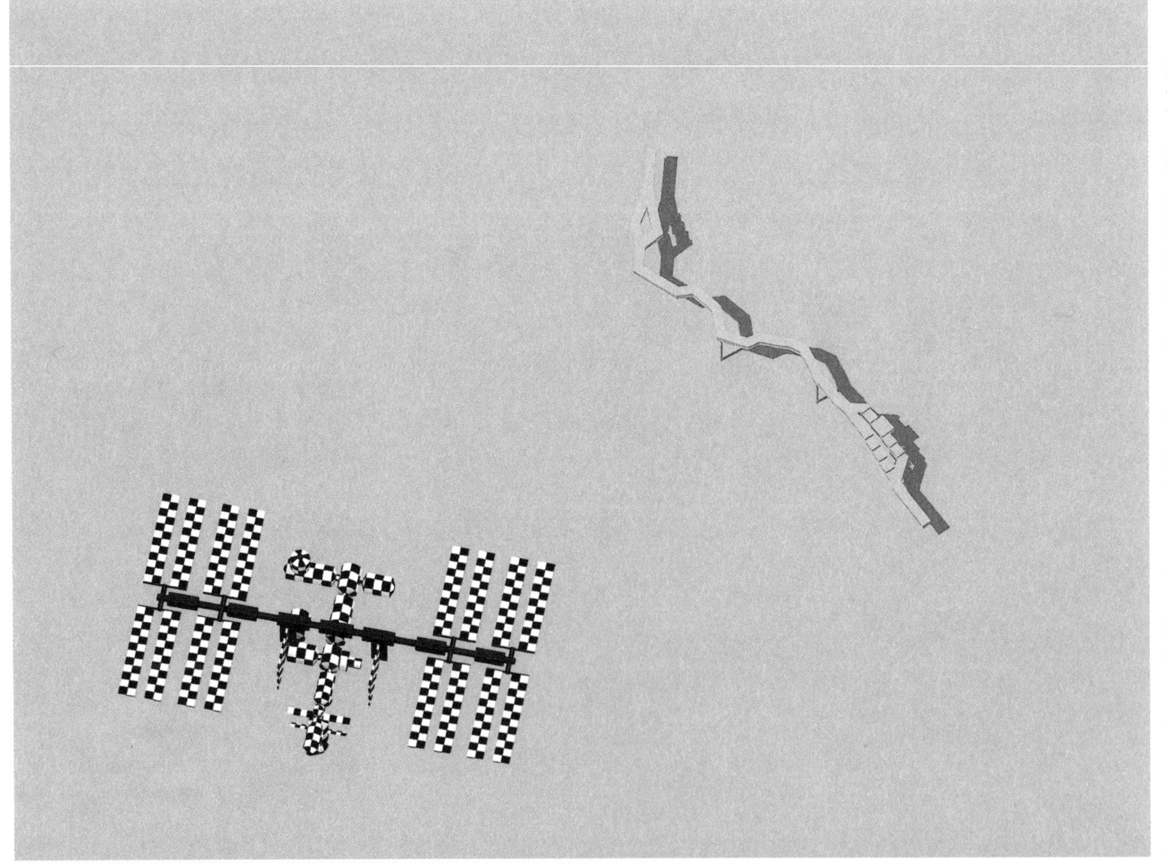

International Space Station. Computer model: Franco Chen. Illustration: Dunne & Raby.

The Chuo Shinkansen Maglev, L0 Series

An object with only one purpose. To be the fastest land-bound object in the world by covering three hundred miles in sixty-seven minutes. Developed to cut down travel times, and "shrink" a country by collapsing time and space.

Ideology: constitutional monarchy with a parliamentary government
Time frame: 2011-2027
Labor force: not known
Cost: US$85 billion (private funding)

Japan GDP: $4.231 trillion (2022)
Japan GPD per capita: $34,135 (2022)

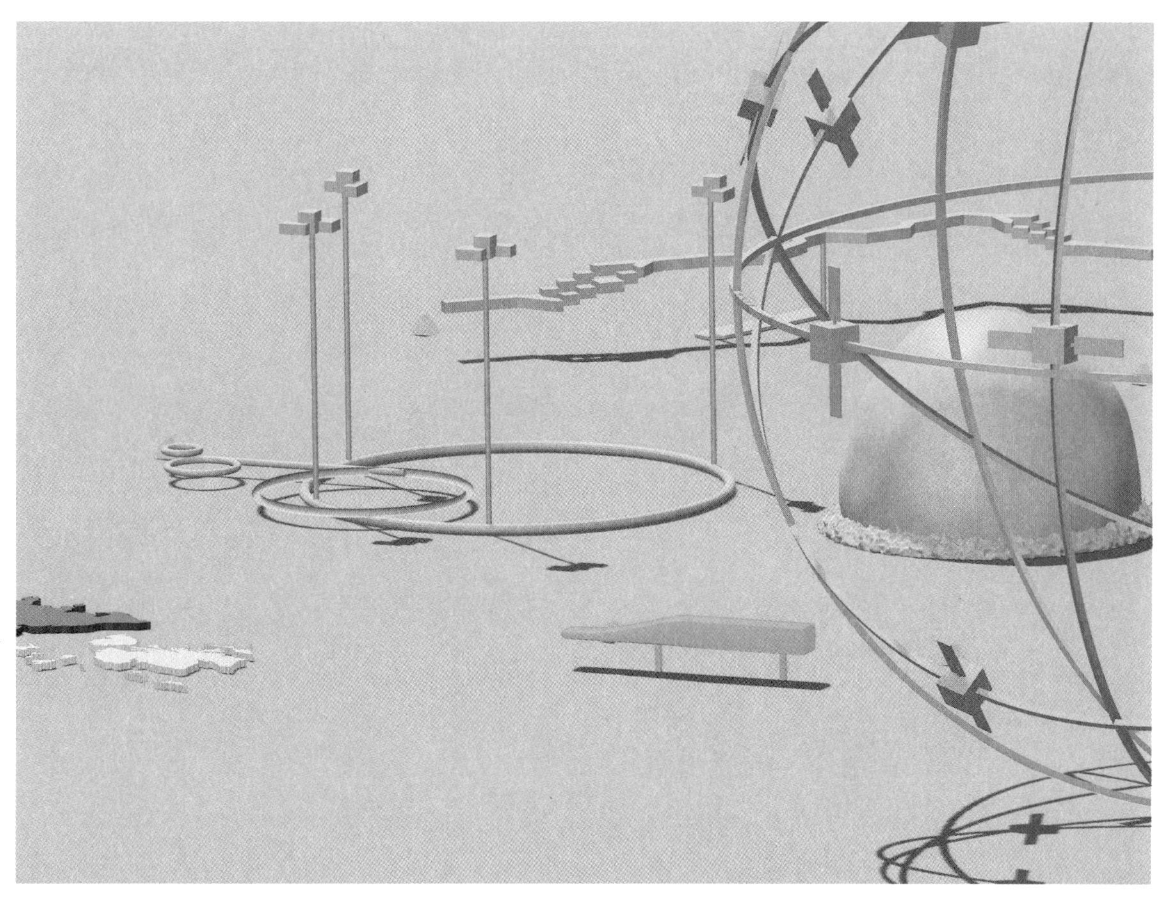

The Chuo Shinkansen Maglev, L0 Series. Computer model: Franco Chen. Illustration: Dunne & Raby.

Trinity Test, the Manhattan Project

To develop an object that would end a war. The redirection of scientific discovery toward an almost unimaginably dark end that even today, still risks ending all life on the planet. But for us, it is the momentary object created in 1945 on the Trinity test site that we include here—"a plutonium device, called 'Gadget,' detonated at precisely 5:30 am over the New Mexico desert, releasing 18.6 kilotons of power."[12] Vast, and fleeting, yet it reshaped the collective imagination and the kind of reality to emerge from it.

Ideology: Anglo-American war effort
Time frame: 1942–1947
Labor force: 130,000 (at peak)
Cost: US$2.2 billion (1945),
US$22.8 billion (2021)

US GDP: $25.463 trillion (2022)
US GDP per capita: $75,269 (2022)

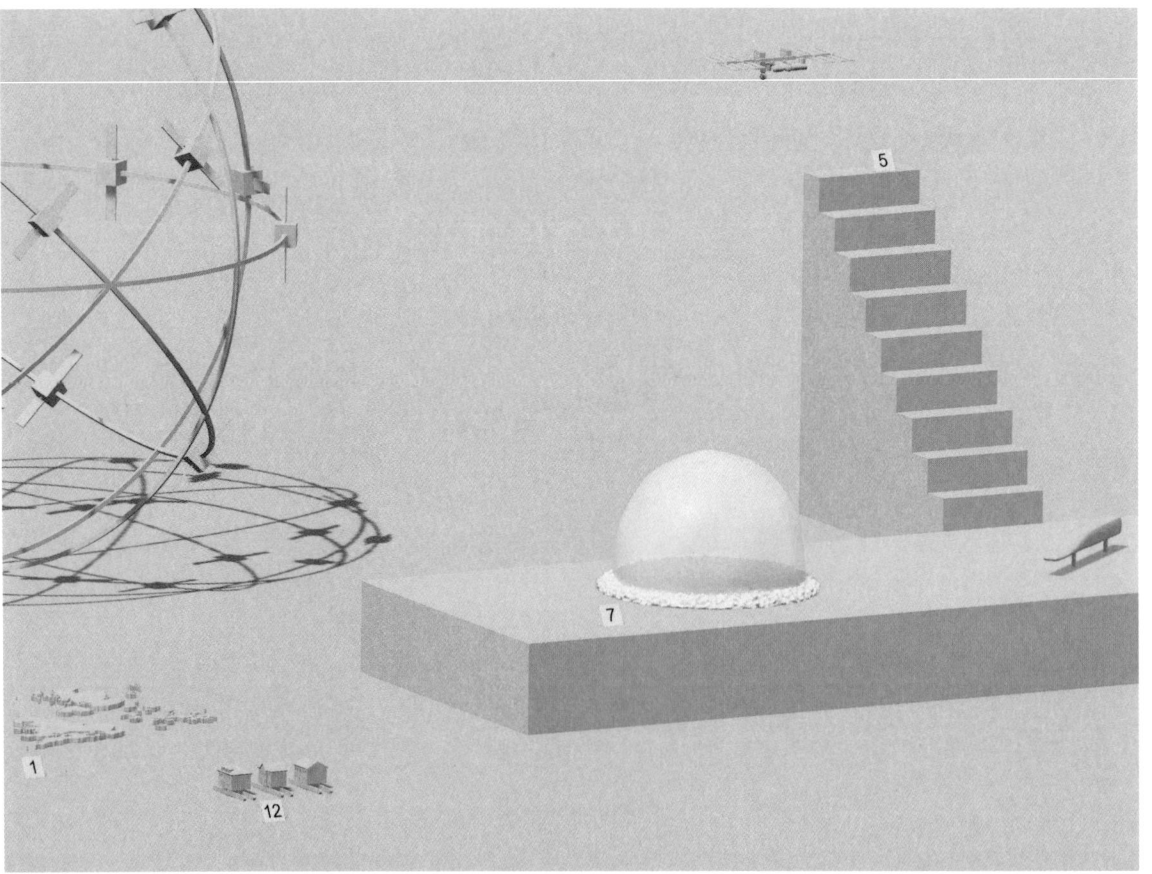

Trinity Test, the Manhattan Project. Computer model: Franco Chen. Illustration: Dunne & Raby.

Panama Canal

The creation of a very large (negative) object through the removal of matter—the most in history. Another project attempting to shrink a little pocket of space-time, cutting the journey from the West to East Coast from twelve to five thousand miles by linking two separate oceans. A heroic project for which many workers sacrificed their lives.

Ideology: Theodore Roosevelt era US liberalism
Time frame: 1870-1913
Labor force: 5,609 lives lost (unofficially 20,000+)
Cost: US$375-639 million (1914),
US$14.3 billion (2007)

Panama GDP: $76.52 billion (2022)
Panama GDP per capita: $17,358 (2022)

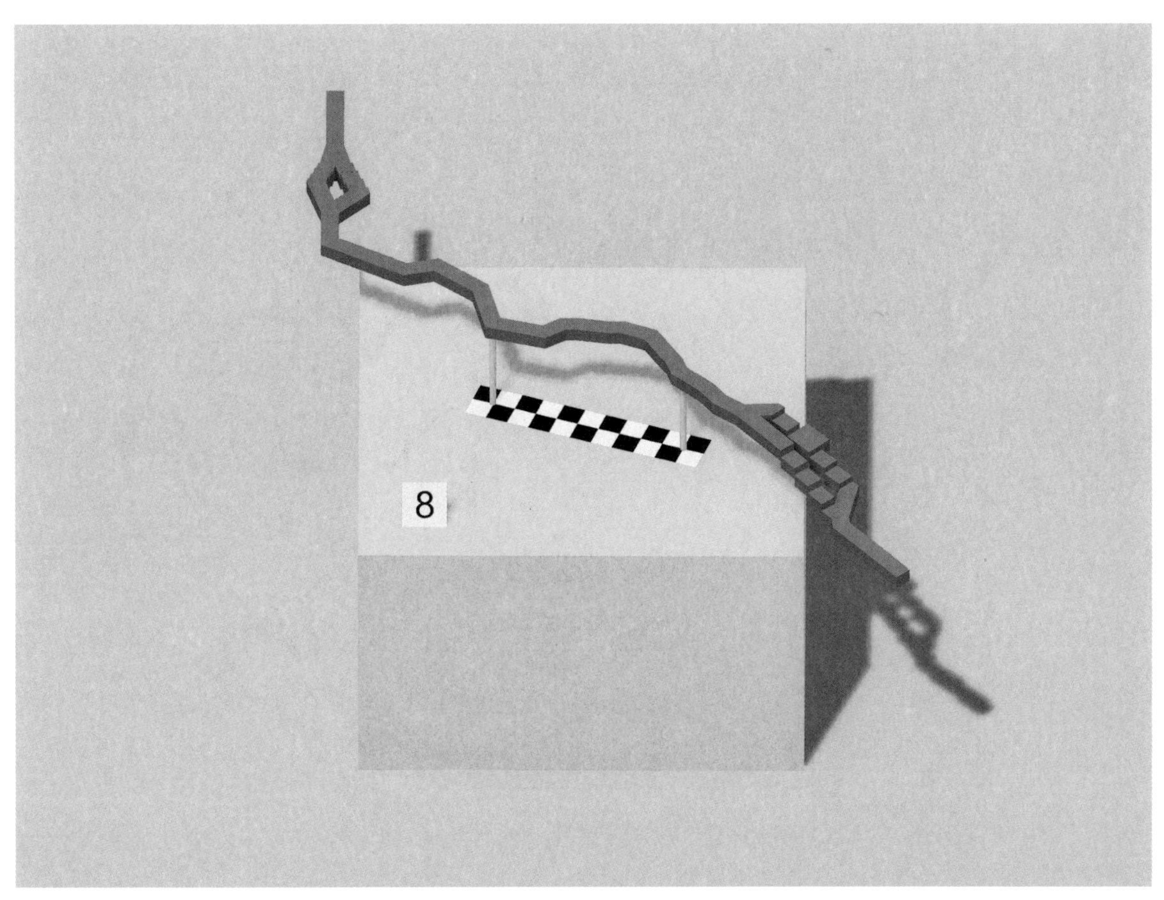

Panama Canal. Computer model: Franco Chen. Illustration: Dunne & Raby.

GPS (24 Satellites)

A very large object made from twenty-four smaller objects. An invisible canopy that encases the world, requiring Einstein's theory of general relativity for it to be realized. Clocks run faster at a 20,200-kilometer altitude. Everything can have a precise position in the world. And what a world it has given rise to.

Ideology: space race
Time frame: 1958 > 1995 (went public)
Labor force: not known
Cost: US$12 billion + US$750 million per year (2012)

US GDP: $25.463 trillion (2022)
US GDP per capita: $75,269 (2022)

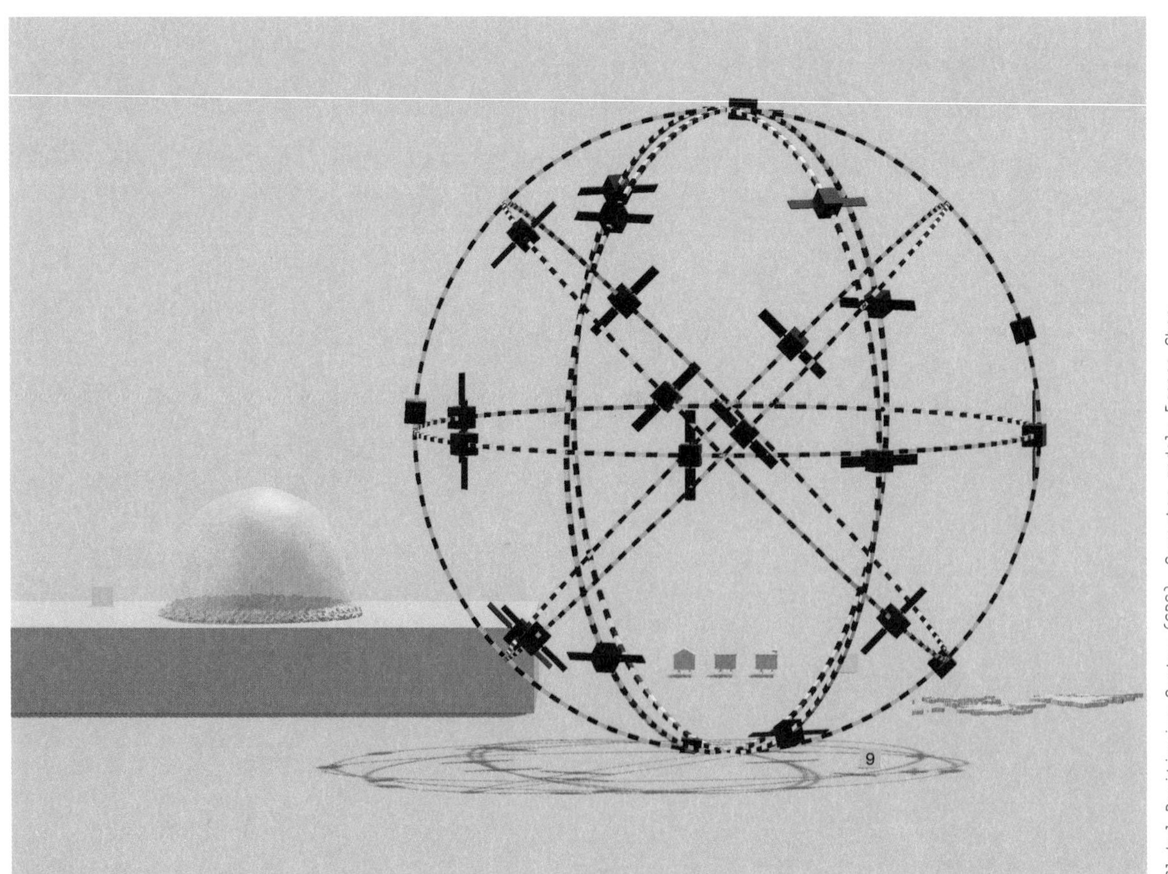

Global Positioning System (GPS). Computer model: Franco Chen. Illustration: Dunne & Raby.

Large Hadron Collider

Another genuinely very large object. The world's largest and highest-energy particle collider that proved the existence of the Higgs boson. A huge circular tube 16.565 miles circumference (5.2 miles diameter). The most recent installment of a postwar dream to bring scientists and nations together. As much a diplomatic project as a science project. The interior of the collider is the coldest place on earth and has the lowest vacuum pressure on earth. A tubular-shaped world within a world.

Ideology: internationalism
Time frame: 1998-2008
Labor force: 10,000 scientists collaborated
Cost: US$9 billion

Switzerland GDP: $808 billion (2022)
Switzerland GDP per capita: $92,410 (2022)

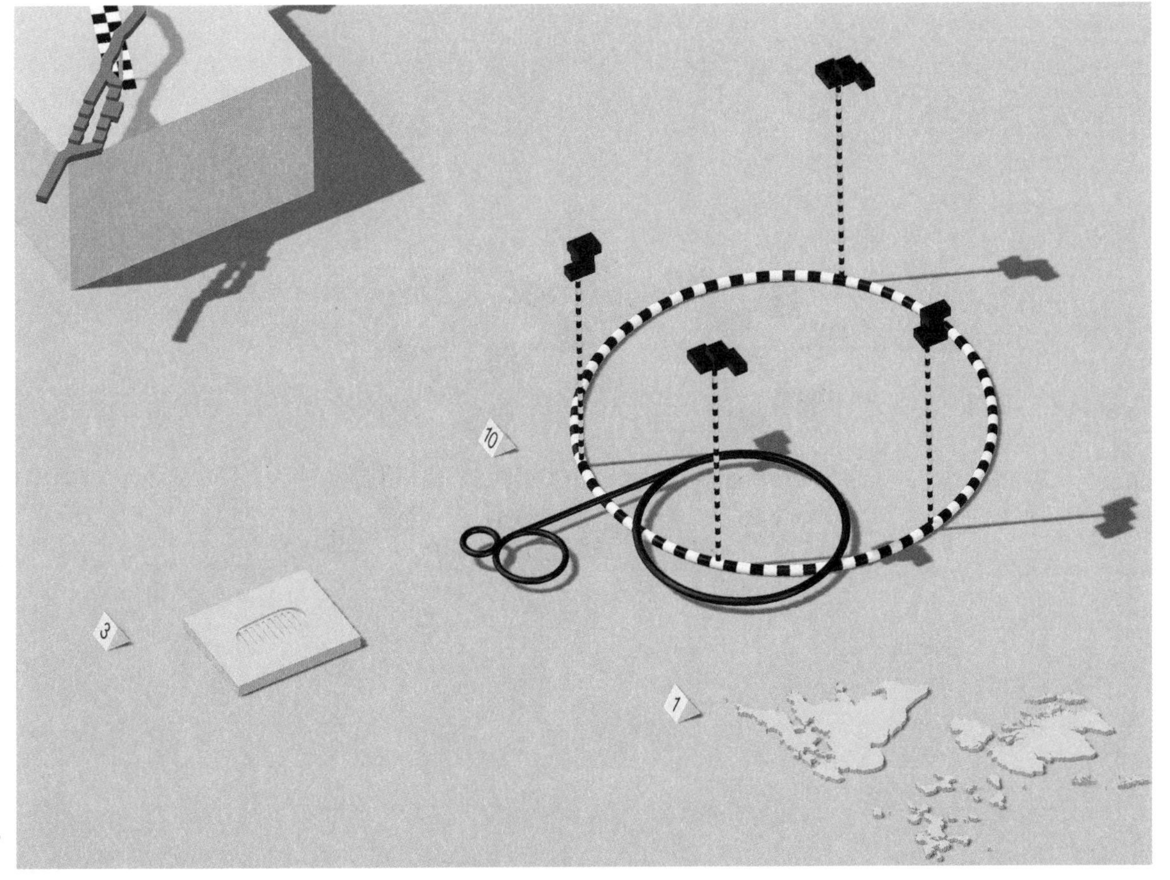

Large Hadron Collider (LHC). Computer model: Franco Chen. Illustration: Dunne & Raby.

B-21 Bomber

An impossible object in so many ways. Defying the laws of aerodynamics, able to change its electronic form at will. Its purpose is to symbolize and materialize a threat. To exude a level of technological awe that renders the possibility of an enemy attack impossible. Almost. One hundred will be built.

Ideology: post-Cold War
Time frame: 2011-2027
Labor force: not known
Cost: US$750 million each, 100+ planned

US GDP: $25.463 trillion (2022)
US GDP per capita: $75,269 (2022)

B-21. Computer model: Franco Chen. Illustration: Dunne & Raby.

Kiruna

Moving a city center consisting of 450,000 square meters of buildings, 3 kilometers (1.8 miles). Not such a large object in terms of budget, but interesting and quirky. The aim here is not to create a new very large object but simply move an already existing one, a city center. For a while, a number of buildings had two addresses: where it is now, and where it either was or will be.

Ideology: social democracy
Time frame: 2014-2035
Labor force: not known (population 22,906; 6,000 will be relocated)
Cost: US$1.2 billion

Sweden GDP: $586 billion (2022)
Sweden GDP per capita: $55.543 (2022)

Kiruna. Computer model: Franco Chen. Illustration: Dunne & Raby.

8 A Nonst Incomplet Glossary Not Here,

andard,
e
of the
Not Now

Chapter 8

A Nonstandard, Incomplete Glossary of the Not Here, Not Now

ABSTRACTA

A relatively modern distinction between entities and objects made entirely from concepts. Abstracta include numbers, ideas, mathematical objects, laws, and money—generally things that do not exist in space and time, or the here and now.

ABSURDISM

The belief that a search for meaning is pointless as there is only purposelessness in the world. The cultural expression of the conflict, contradictions, and paradoxes generated by such a search. In this context, absurdity can serve as a form of defamiliarization, putting a design proposition constructively at odds with the here and now, drawing a viewer into a space of reflection.

AFROFUTURISM

A cultural expression of the meeting of science fiction and African American culture (or Black diaspora), history and struggles. Most people agree the term was first used by Mark Dery in his "Black to the Future" essay in 1993. It offers a radically different kind of imagination to the limited tropes and visual clichés of a Western science-fictional imagination. Far removed from the here and now. See writer Octavia E. Butler's *Lilith's Brood* collection for an example. Different from African futurism, which is rooted in native African culture and history.

ALTERNATIVE

Used as a prefix, it can sound a little quaint, signaling a more experimental form of design, for example. But we use it more to suggest other ways of being—alternatives to how things are now, here. By suggesting an alternative to the current status quo, it embodies an implicit critique. This is where it all starts and ends. A desire for alternatives, not to impose or prescribe, but to prompt, inspire, and activate other imaginations. Alternatives that are generative rather than destinations.

ANALOGICAL SCIENCE FICTION

Darko Suvin's alternative to extrapolation as a basis forspeculation.[1] Whereas extrapolation usually starts with the here and now, an analogical approach begins with ideas. Often blends concepts from philosophy with unconventional approaches to world-building. Examples include stories by Jorge Luis Borges and Edwin Abbott Abbott.

ANTI-ANTI-UTOPIANISM/ NEGATIVE UTOPIANISM

An idea informed by philosopher Fredric Jameson's reflections on the value and contradictions of utopian thought. Not believing in a specific utopia as such, while believing that there must be better alternatives to the present situation, even if currently unknown. A belief that it is not enough to say that what we have now, the here and now, with all of its faults, is the only possibility and we should stick with it simply because to think otherwise is "unrealistic." A critique of a lack of imagination rather than of the utopian impulse.

AS IF

In design, rather than asking, "What if...?" one can design "as if" something has already happened or is true, bringing the not here, not now into contact with the here and now.

AUXILIARY REALITY ZONE

Part of reality, but also separate from it. Operating at or just beyond the edges of normal conventions of logic, economic sense, utility, and practicality, auxiliary reality zones could serve as catalysts for collectively exploring and imagining alternative ways of seeing and being in the world, materialized through the stuff of everyday life. Places where design and philosophy meet to provide not solutions but instead new ways of thinking about the designed world. Realists could, of course, dismiss this as a form of escapism, but the value of a place like this is in expanding people's imaginative horizons, just a little, in a pleasurable way, supplying cultural nourishment and a new "climate of possibility." This is valuable in itself as a sort of civic act that stimulates the public capacity for imagination by sustaining and enriching the worlds that people carry around inside them, from which new realities emerge.

BEING

Whatever *is*. Having existence. Which is not as straightforward as it might seem. Much has been written about what is or exists. Some philosophers claim that even nonexistent things exist, conceptually, by virtue of being able to say "a thing that does not exist." Where things that do not exist, exist, is another matter.

CHUKJIBEOP AND BIHAENGSUL

Inspiration for the title of Korean artist duo Moon Kyungwon and Jeon Joonho's contribution to the 2015 Venice Biennial, "The Ways of Folding Space & Flying." Based on Taoist practice, Chukjibeop means a "hypothetical method of folding space and of allowing one to travel a substantial distance in a short space of time."[2] Basically, a form of teleportation or "earth shrinking." Bihaengsul means to levitate or fly. While acknowledging these concepts might appear ludicrous, the artists write that for them, these terms represent the power of the imagination to overcome human limitations. For us, they collapse spatiotemporal distinctions between here and there, and now and then.

COGNITIVE ESTRANGEMENT

For Darko Suvin, this is the most important purpose of science fiction. To invent worlds that estrange the reader from their own empirical reality so they can critically reflect on the way they live. Introduced in his book *Metamorphoses of Science Fiction* (1977). A variation ondefamiliarization, the alienation effect (A Effect), and estrangement, tailored to science fiction. Essential for maintaining productive distance between the here and now, and the not here, not now.

COLLAPSE OF THE WAVE FUNCTION

In quantum mechanics, the moment when multiple possibilities in a superposition collapse into one due to interaction with the external world when it is "observed" (or measured). Hence the huge efforts that go into isolating quantum computers from their environment or reality. Not valid for the MWI, where multiple worlds branch off intact rather than collapsing into each other. An idea ripe for absorption into everyday language as we move away from a situation where Newtonian concepts help make sense of current realities.

CONSENSUS REALITY

An agreed-on reality—or at least its underlying concepts—in any group, community, or society. The gravitational pull that crushes efforts to reach a not here, not now, even in the imagination, at least in design.

CORRELATIONISM

As Steven Shaviro puts it in *Extreme Fabulations* when writing about Quentin Meillassoux, "For the correlationist reality can never be separated from our projections upon it; we only encounter phenomena. Meillassoux laments that, in the wake of what he calls the 'Kantian catastrophe,' we are cut off from the 'great outdoors' of 'absolute reality,' and trapped within the narrow circle of our own all-too-human constructions. In this world of mere phenomena, our telescopes and microscopes do nothing more than reflect our own presuppositions back to us. Ontology (the study of the way things actually are) is ruled out of bounds by Kant and his successors, and replaced by phenomenology (the study of the way things appear to us) and epistemology (the study of how we are able to know the things we know)."[3] With this understanding comes the sad realization that something lies beyond the not here, not now, and is forever inaccessible to our human-shaped imaginations.

COSMOTECHNICS

In the words of Yuk Hui, who coined the term, "The unification between the moral order and the cosmic order through technical activities."[4] A call for a plurality of technological imaginations, for new ways of thinking, imagining, and world making as an alternative to technological reason. As Hui puts it in an interview on Taoist robots, among other things, "Because our technological creations are challenging historical limits through climate change, artificial intelligence and synthetic biology, it is critical to reexamine the diversity of cosmotechnics, or how technology is infused with a worldview."[5]

COUNTERFACTUAL
Reshuffling the here and now. What might have happened if...Often used to provide new perspectives on present realities by revisiting points in history when a different decision could have been made, resulting in a radically different version of the here and now. Literally means "contrary to the facts." Rarely used in design, yet has much critical potential.

COUNTERFUTURISMS
Culturally rooted forms of futurism (especially in literature, cinema, and fine art) that offer alternatives to a universalist approach to the not here, not now. They often embody or reflect radically different values, belief systems, cosmologies, worldviews, and sometimes ontologies. Probably the best known is Sinofuturism, which was originally developed by Western observers, but later embraced and put to new uses by Chinese writers. For example, Liu Cixin's *The Three-Body Problem*, translated by Ken Liu (2014). Other kinds of counterfuturisms include African futurism, Afrofuturism, Arab futurism, Black quantum futurism, Gulf futurism, Chicanafuturism, Islamofuturism, Sufi futurism, Hindufuturism, Buddhist futurism, Sikh futurism, Judeofuturism, Taoist futurism, Confucian futurism, and Shamanic futurism.[6]

DELUSION
When a reality is mistaken for a fiction. Or when the not here, not now, and the here and now, swap places.

DEMOCRACY OF OBJECTS
A variation on object-oriented ontology put forward by philosopher Levi Bryant: "The democracy of objects is not a political thesis to the effect that all objects ought to be treated equally or that all objects ought to participate in human affairs. The democracy of objects is the ontological thesis that all objects, as Ian Bogost has so nicely put it, equally exist while they do not exist equally. The claim that all objects equally exist is the claim that no object can be treated as constructed by another object. The claim that objects do not exist equally is the claim that objects contribute to collectives or assemblages to a greater and lesser degree. In short, no object such as the subject or culture is the ground of all others. As such, The Democracy of Objects attempts to think the being of objects

unshackled from the gaze of humans in their being for-themselves."[7] An imaginative and intellectual challenge for designers of the not here, not now to grapple with.

DESIGN
Today especially, design can mean many things. It evolved to concretize the here and now, but for us, it is about bringing the not here, not now into conversation with the here and now through the design of everyday objects embodying alternative values, beliefs, and possibly ontologies. Materially located here, but belonging elsewhere conceptually.

DIEGETIC PROTOTYPE
An idea put forward by David Kirby in 2010. Describes an object that belongs to a specific future or fictional world. Usually associated with science fiction cinema and the kind of objects developed in design fiction. For us, the bond between object and fictional world it necessitates is a little too tight as one must grasp its world in order to fully appreciate the object, and vice versa. The fictional objects in this book work in a looser way, ideally generating worlds in the mind of the viewer. They hint at worlds rather than conveying them.

DOMESTICATING THE IMPOSSIBLE
H. G. Wells once wrote that to help the reader, the writer should quickly "domesticate the impossible hypothesis"[8] a story is based on. That way, the reader can accept it, leave it behind, and move smoothly into the story by connecting to its characters' very human responses to the seemingly strange situation. His aim was to hold the reader "by art and illusion and not by proof of argument."[9] In the context of the kind of design we explore in this book, this suggests a move away from extrapolation and science as the primary ways of grounding design speculations to poetry and affect.

DREAM
A specific kind of not here, not now. A difficult idea to discuss in design at a time when realism in all of its varieties dominates.

DYSTOPIA
A form of not here, not now based on negative possibilities. Sometimes used as a cautionary tale, especially when extrapolating from the present to the highlight potentially negative consequences of a current path. Less effective when used to entertain, attract attention, or promote oneself.

EERIE
For Mark Fisher, the eerie is a relative of weirdness associated with places, ruins, abandoned buildings, towns, landscapes, and especially absences rather than presences (the weird). It is what is missing that renders a place eerie. People, life, explanation, or meaning.

The feeling of stumbling on lost worlds. An overlooked quality that could be used to produce new forms of constructive estrangement in design.

EPISTEME
An underlying code that structures the way people make sense of the world. Raw material for philosophical forms of world-building.

EPISTEMOLOGICAL REALISM
An idea at the heart of arguments in quantum mechanics about whether it describes the world as it is, or is simply a useful and extremely accurate set of tools for working at a tiny scale. According to philosopher Levi Bryant, "Epistemological realism argues that our representations and language are accurate mirrors of the world as it actually is, regardless of whether or not we exist. It seeks to distinguish between true representations and phantasms."[10]

ESCAPE
Often used as a criticism of imaginative work located in a not here, not now. Something writers of speculative and fantasy fiction have given much thought to. As Ursula K. Le Guin once remarked, "The direction of escape is toward freedom. So what is 'escapism' an accusation of?"[11] Or as J. R. R. Tolkien put it, "In using escape in this way the critics have chosen the wrong word, and, what is more, they are confusing, not always by sincere error, the Escape of the Prisoner with the Flight of the Deserter."[12]

ESTRANGEMENT
A process of defamiliarizing the familiar so it can be seen with fresh eyes, either with renewed appreciation or critique. Strangeness for its own sake suggests novelty as an end in itself—strange for the sake of being strange. Estrangement suggests a process or experience where strangeness is a means of defamiliarizing everyday mindsets and ways of seeing the world. Estrangement, and related ideas such as defamiliarization in literature, the A Effect in theater, and cognitive estrangement in science fiction, are difficult ideas to accept in design as almost by definition, design is intended to fit existing reality, to enter the world seamlessly, strengthening the reality principle and buttressing the here and now.

EUTOPIA
The good place.

EXTRAPOLATION
Projecting into the future based on the here and now to make it more plausibe or realistic. But this also limits its usefulness in expanding people's imaginative horizons as it is still a version of existing reality—a reality that for many people is broken.

So why extrapolate from this position? Well, it is useful for anticipating a possible future or generating cautionary tales. Sometimes criticized for being too instrumental and logical. Perhaps instead we need entirely new starting points, located in a not here, not now.

EXTRO-SCIENCE FICTION
A form of fiction proposed by Quentin Meillassoux "where science is impossible." An example he gives is Rene Barjavel's 1943 science fiction novel *Ravage*, in which electricity stops working overnight, and although several characters speculate on why, no explanation is ever given. Science is not used to explain away a strange anomaly, as it might in a classic science fiction story, but there are still different expectations from fantasy and magic realism too.

FACTIALITY
Quentin Meillassoux's term for a situation where what we think of as "reality," or the realm of phenomena, is contingent and might in time be replaced. In other words, reality could be otherwise. This is due to it being constructed from human concepts grounded in sensations, and can always be improved on.

FAIRY TALE
A step too far.

FAKE
A thing that is knowingly claiming to be something it is not. Different from "fictional," which makes no attempt to pretend it is real. A false reality presented as something authentic. Fictional realities declare their unreality at all times. The material fictions we pursue signal their fictional status through their aesthetic treatment. They are fictions presented as fiction. No attempt is made to trick a person into suspending disbelief; it always takes the form of an invitation.

FANTASTICAL
Something that has an excessive amount of unreality to an often quite enjoyable and surprising extent.

FANTASY
A debased term. Difficult to use as it almost always attracts a negative response in design. Usually conjuring up associations with indulgence, irrelevance, and escapism. Associated with hobbits, elves, and magic. A not here, not now we stay away from.

FICTA
Fictional entities including people, animals, monsters, places, and objects. Both here and not here.

FICTION
Something made up. To entertain, prepare, warn, inspire, set a path, stretch the imagination,

or provoke. The imaginative engine of the not here, not now. Also, numbers, laws, ideas, nations, money, concepts, and so on. A notion some claim makes us uniquely human.

FICTIONAL REALISTS
Those who believe there are entities that have the property of not existing in some way. For example, there are objects that do not exist, yet are still objects.[13]

FRACTIVERSE
A term used by sociologist and science and technology studies scholar John Law when discussing alternatives to a "one-world world" and the challenges of the pluriverse: "And here, in turn, is why this is politically important. If we live in a single Northern container-world, within a universe, then we might imagine a liberal way of handling the power-saturated encounters between different kinds of people. But if we live, instead, in a multiple world of different enactments, if we participate in a fractiverse, then there will be, there can be, no overarching logic or liberal institutions to mediate between the different realities. There is no 'overarching.' Instead there are contingent, local and practical engagements. The implication is that if we are to work well in postcolonial encounters we will need to craft ways of doing so that are themselves contingent, modest, practical, and thoroughly down-to-earth; ways of proceeding that acknowledge and respect difference as something that cannot be included. So that is the problem. Whether to assume the world is one and we're all inside it. Or to wrestle with the implications that worlds in the plural are enacted in different and power-saturated practices."[14] A challenge rarely acknowledged in discussions to do with designing for a pluriverse.

FULL EMPTY
An object, in the brothers Strugatsky's story *Roadside Picnic*, made from words. One of several mysterious and possibly alien technologies discovered in the "Zone." It consists of two parallel disks held about four hundred millimeters. apart by an unknown force. Between them is an incredibly heavy blue liquid. Most examples are empties and simply consist of two disks held in parallel. Objects can be passed through the space between the disks.[15] A possible addition to Meinong's taxonomy of objects, somewhere between round squares and perpetual motion machines.

FUTURE
An overused device for framing the not here, not now in design. Once a project steps away from existing reality it is usually relocated here. A somewhat limited form of speculation that ties "elsewheres" to present realities through extrapolative threads.

There are many other ways of framing speculative work yet to be explored.

GENERATIVE ARTIFICIAL INTELLIGENCE TECHNOLOGIES

The use of neural nets to generate images from text prompts. Sometimes called artificial intelligence art. When handled well, a helpful tool for leaving behind the here and now, although because it feeds on the already imagined, its images can feel a little familiar. Will be most interesting if, or when, it can work with underlying concepts and logic, not just appearances, to generate objects that exceed the human imagination in some way—what we currently think of as humanly impossible objects.

GREAT OUTDOORS

According to Steven Shaviro, for Quentin Meillassoux, "The 'great outdoors' of things in themselves is 'an absolute that is at once external to thought and in itself devoid of all subjectivity' as of all life: a materiality that is 'dead through and through' ('Iteration, Reiteration')."[16] An inaccessible space lying beyond the human umwelt or human sensory and cognitive world that we cannot ever know. Can design point to such places?

HETEROTOPIA

A kind of reality that does not belong in the present or the here and now. Yet here it is, in front of us, physically present, but conceptually elsewhere, fusing the here and there with the now and then.

HOAX

A (poor) form of fiction designed to fool or trick people into believing it is real through the use of realist aesthetics. Raises ethical issues in relation to design fiction and speculative forms of design practice. Could be considered "too easy" as a strategy or a form of cheating as it doesn't invite the viewer to suspend their disbelief but does it for them. Something relocated seamlessly from the not here, not now into existing reality. Abandons conventions associated with signaling something is in fact fiction, thereby defeating the purpose of designing for the not here, not now.

HYPEROBJECTS

A word coined by Timothy Morton in 2008 "to describe all kinds of things that you can study and think about and compute, but that are not so easy to see directly: hyperobjects. Things like: not just a Styrofoam cup or two, but all the Styrofoam on Earth, ever." Objects that are just too big for a human-sized imagination.[17]

HYPERREALITY

A once intriguing concept that now seems rather quaint in relation to the onslaught of different degrees of reality people experience on a daily basis. Originally used by sociologist and media theorist Jean Baudrillard to mean a simulation that appeared super real, or more than real—hyperreal. Something existing here and now, but through an excess of realism, ceases to be a representation of something, or even a simulation, and becomes a new reality itself—a hyperreality.

HYPERSTITION
A self-fulfilling prophecy. Coined by the Cybernetic Cultures Research Unit associated with the philosophy department at the University of Warwick in the 1990s. As Nick Land puts it, "Superstitions are merely false beliefs, but hyperstitions—by their very existence as ideas—function causally to bring about their own reality. Capitalist economics is extremely sensitive to hyperstition, where confidence acts as an effective tonic, and inversely. The (fictional) idea of Cyberspace contributed to the influx of investment that rapidly converted it into a technosocial reality."[18] Sometimes used to justify moralistic critiques focused on accountability in science fiction, design speculations, and designing for the not here, not now.

IDEA
A raw material to work with.

IDEALISM
The opposite of materialism (where only physical things exist). In idealism, reality is a product of the mind where only the mind exists. Which naturally gives rise to all sorts quandaries of which much has been written. Two key approaches are "ontological idealism" (e.g., George Berkeley), where reality is based on ideas or the mind, and all that exist are ideas; and "epistemological idealism" (e.g., Immanuel Kant), where although things or reality can exist independently of the mind, we can only know it through the mind, so our knowledge of it is still limited to how the mind works. The properties we associate with a thing are not of the thing itself necessarily but rather a by-product of how the human mind makes sense of the world, things, and reality itself.[19] Interesting in this context as a starting point for ontological speculation in world-building. See, for example, Jorge Luis Borges's "Tlön, Uqbar, Orbis Tertius."

IDEOLOGY
A carefully crafted lens, made from ideas, for viewing the here and now in a specific way. Can be illuminated through designs for a not here, not now. More technical and precise than a worldview.

ILLUSION
A fiction mistaken by a viewer for reality. When the not here, not now is mistaken for the here and now.

IMAGINATION
An endangered and precious resource. A neglected public good. A political battleground. Home of the otherwise. A place where possible worlds are found, built, and inhabited.

IMPOSSIBLE
A forever shifting, unstable boundary between the real and the unreal. Between what is and what could be, lying just beyond the limits of understanding and imagination. Something that belongs to a different reality system than the prevailing one.

IMPOSSIBLE OBJECT
Impossible objects are seemingly incompatible with existing sensemaking frameworks, or at least they make a different kind of sense. But they are not nonsense objects or Dada-like gestures challenging the very notion of sense. This is why they are valuable; they are invitations to question the frameworks we use to make sense of the world and even what the "world" is. Objects for exploring other systems of thought that stretch the imagination in ways that not only help imagine how things could be otherwise but also how the way we think about the metaphysical work in progress we call reality might change too. Impossible in this context means within our own narrow sense of what we deem real or unreal. What can and cannot exist within our specific ontology.

IMPOSSIBLE WORLDS
In philosophy, "impossible worlds often characterized as ways things could not have been." And "another definition has it that impossible worlds are worlds where the laws of logic fail." And yet "another definition has it that impossible worlds are worlds where the laws of *classical* logic fail."[20] An extreme form of not here, not now designed around contradictions and paradoxes that tests the basis of human thought and ideas.

IMPOSSIBILIA
Technically, "the objects exemplifying absolute impossibilities which inhabit impossible worlds."[21] For us, a collection of impossibilities richly materialized in a variety of forms in order to study, compare, and reflect on.

INEFFABLE
An idea too great or extreme to be expressed in words (dictionary definition). Conceptually located just beyond the boundaries of human thought and imagination, generating a quandary: how can we know something unknowable without saying it and therefore making it knowable. This truly impossible object is not one we can ever encounter, at least in design, and is best left to philosophers, although it does usefully mark one end of a spectrum.

INNER WORLDS

The worlds we carry around inside us, unique, in theory at least. Often productively at odds with the world out there. Usually neglected by design where the focus is typically on external worlds. A precious place where ideas form and new realities begin to emerge. Under threat from mass homogenization. Some believe it is the worlds we carry around inside us that need to change first, if the world out there is to eventually change for the better.

INSOLUBIA

A collection of unexplainable problem figures related to "impossible figures."

IRREALISM

Related to the unreal, but slightly different. More concerned with an unknowable real, just beyond human-centric notions of reality. Suggests something a little more strange than the unreal or antireal. There are several literary, fine art, and philosophical variations, but it is best captured in this description: "as much as we might try to order our world with a certain set of norms and goals (which we consider our real world), the paradox of a finite consciousness in an infinite universe creates a zone of irreality ('that which is beyond the real') that offsets, opposes, or threatens the real world of the human subject."[22]

LARGER REALITY

A take on reality articulated by Ursula K. Le Guin in her 2014 National Book Awards speech, where she described writers and artists as "realists of a larger reality."[23] For us, it suggests a reality that contains ideas, dreams, hopes, abstracta, and impossibilities usually banished to unreality. We live among these ideas and thoughts, and they are part of reality, just in a different way.

LAWS OF PHYSICS

A description of the outer envelope of the human umwelt.

LIFEWORLDS

A concept sometimes used in the social sciences to mean a sense of the world that is grounded in lived experience and what is presented to the senses, here, now. Philosopher Edmund Husserl is credited with its popularization.

MAGIC REALISM

A form of fiction rarely encountered in design. In literature, one change, but a dramatic one, is made to existing reality. For example, José Saramago's *The Stone Raft*, where the Iberian Peninsula breaks off from Europe and slowly drifts into the Atlantic Ocean. Could this allow for new settings to design within, transforming the here and now into a strangely familiar not here, not now?

MANY WORLDS INTERPRETATION (MWI)
A highly controversial interpretation of quantum mechanics put forward by Hugh Everett in 1957 and popularized by Bryce DeWitt in the 1970s. For its advocates, it is one of the simpler and more elegant interpretations requiring the least additional elements by taking quantum theory literally, as a description of reality. According to the MWI, wave function collapse never happens, so multiple possible states do not collapse into one state in the observer's timeline but instead continue to exist in many other parallel timelines created at that moment. Its critics argue that it is ontologically extravagant. Where, for example, do all of these worlds or universes that are continuously being created exist? It is understandably a popular idea in fiction and has led to many enjoyable speculative tales. David Deutsch, one of the original conceptualizers of quantum computing, goes as far as to suggest that this is what gives quantum computers their edge: they are literally calculating in multiple worlds simultaneously.

MATTER
A contradictory physical presence made primarily of space, not stuff. What we feel when we touch a matter-ial object are the forces between electrons. Something we become aware of when we look at images made using X-rays, reminding us that physical reality is actually not as solid as it might feel to our touch.

MEINONGIANISM
Philosopher Alexius Meinong thought that "over and above the concrete entities that exist spatiotemporally [e.g., mountains] and the ideal or abstract entities that exist non-spatiotemporally [e.g., numbers], there are entities that neither exist spatiotemporally nor exist non-spatiotemporally [e.g., round squares]: these are the paradigmatic Meinongian objects that lack any kind of being."[24] He once said, "There are objects such that it is true to say of them that there are no such objects."[25] Understandably, his ideas are quite controversial.

MEINONGIAN OBJECTS
Designers say there are real and not real objects, and they want to deal with the real. Meinong says there are objects that are real, sort of real, hyperreal, not quite real, really real, and so on. We want to design for this world. A world that more fully reflects the range of objects that people interact with daily.

MEINONG'S JUNGLE
The place where objects like this can be found. Intended as a derogatory term to suggest Meinong's theory of objects is a conceptual mess that muddles important distinctions between different kinds of being.

It appeals to us, though, and feels like a useful starting point for attempting to navigate today's multimodal, layered realities.

METAPHYSICS
When used in everyday life it usually means something highly abstract, maybe even ungraspable, and certainly not approachable for design. But of course metaphysicians would beg to differ. From an intellectual perspective there is so much of interest—the nature of reality, what lies beyond human perceptions of reality, and how it can be conceptualized. What would design that concerned itself with the metaphysics look like?

MINDSET
An attitude that determines how a person views and reacts to external events. More compact than a worldview. The target of designing for the not now, not here.

MOCK-UP
A premodel of a model for testing some aspect of a model.

MODAL FICTIONALISTS
A philosophical take on the not here, not now. Rather than saying it is possible that a "vegetable lamb" exists (in the here and now), they would say there is a possible world where a "vegetable lamb" exists (not here, not now). Modal fictionalists enjoy the benefits of using possible worlds without believing in them. Similar to useful fictions and thought experiments.

MODAL REALISM
In philosophy, modal realists believe that nonactual possible worlds and other nonactual entities such as Sherlock Holmes are as real as the actual world and its contents, just in a different way. For more on this, see David Lewis's On the Plurality of Worlds.[26]

MODEL
An idea made tangible. Not necessarily intended to be realized, but used as a tool to think with or convey an idea. A useful fiction. By necessity, a model is often stripped down and simplified or abstracted in some way. Different from a prototype, which usually tests an idea against existing realities, whether technical, physical, ergonomic, and so on. The question a prototype answers is always, Does it work? Whereas models offer a little more freedom and openness. They are never complete, and can only ever partially represent an idea or thought, inviting a more nuanced kind of engagement from those who encounter them.

MULTIVERSE
Probably one of the most popular and entertaining quantum-related tropes in literature and cinema. Also used as a kind of business tool in some science fiction

franchises and cinematic universes to allow previously killed-off superheroes to return.

MYTHOPOESIS
Mythmaking in literary and cinematic fiction. A form of conceptual and poetic world-building less informed by science.[27] We need more of this in design.

NAIVE REALISM
A belief that the world is as we perceive it. That the qualities we see in an object, belong to that object, rather than resulting from the unique workings of the human senses and mind.

NATURAL
What we expect to happen according to the laws of nature. Related to, but different from, preternatural and supernatural.

NATURALISM
A form of representation in fiction aligned with how things are now. Truthful. Accurate. Proper. But a little restrictive aesthetically.

NEW MATERIALISM
A renewed interest in materiality, matter, and processes of materialization viewed through various interconnected philosophical lenses. Matter as something vital, active, more than passive, and embodying social and political ideas. Embracing a shift from anthropocentric to ecocentric thought, where nonhuman entities are viewed as more active in shaping the world than previously thought. Collapses binaries such as nature-culture, material-immaterial, mind-body, human-nonhuman, and so on. See, for example, Karen Barad, Rosi Braidotti, and Jane Bennett.

NEW WEIRD
A cultural genre that weaves weirdness, eeriness, and ecological thought into a form of literary and filmic world-building. See, for example, Jeff VanderMeer's The Southern Reach Trilogy (2014). We need some of this in design too.

NOMOLOGICAL IMPOSSIBILITY
Scientifically impossible as defined by the laws of nature.

NONLOCALITY
The technical term for what Albert Einstein called "spooky action at a distance": the ability of two objects separated by vast distances to know each other's states.[28] An affront to Einstein's understanding of the universe as it meant information must travel faster than the speed of light, which was impossible according to his theory of relativity. Einstein and two colleagues proposed a thought experiment called the Einstein-Podolsky-Rosen paradox to prove quantum mechanics was incomplete and that a hidden variable must exist that once discovered, would resolve the paradox of what is now known as entanglement. Since then it has been proven that there are

no hidden variables and quantum mechanics is indeed spooky. Once entangled, separate objects in two places exist as one—a property known as nonlocality.

NONREALIST
The unrealist knows that they are not being realistic, they are being unreal by design, whereas it's possible that a nonrealist may think they are being realistic. Important when it comes to fiction in design practice, especially certain kinds of speculative solutionism where a design proposal is frequently presented as a "real" solution yet is clearly a fiction. Often fictionalists in denial.

NOUMENAL WORLD
How things actually are (noumena). Although we can never know how things actually are—the noumenal world. Compared to how things appear to us as phenomena.

NOVUM
Latin for new thing. Used by Darko Suvin in relation to science fiction where "the hypothetical 'new thing' a story is about, can be imagined to exist by scientific means rather than by magic, i.e., by the *factual reporting of fictions* and by relating them in a plausible way to reality."[29]

OBJECT
A thing that has become objectified. When we are not observing a thing, it is still there, but not necessarily in a form accessible to humans in its entirety; we might never know what exactly is there. In this sense, what we perceive to be objects are the products of our sensory and cognitive apparatus, and we can never really know what lies beyond this.

OBJECTIVE REALITY
A comfortable fiction.

OBJECT-ORIENTED ONTOLOGY
A philosophical approach that views the nonhuman object as a thing it itself, independent of human thought. Addresses all things in the universe with equal weight. For some (e.g., Graham Harman), objects include ideas, events, energy fields, and so on.

OBJECTUM
A medieval term meaning "thing presented to the mind" used by neo-Meinongian Richard Routley when discussing his view that nonexistent objects, or nonentities, are not nothings.

ONE-WORLD WORLD
Social scientist John Law's term for a Western liberal world that contains all others. Other communities can have different perspectives or views on this world, but not their own worlds.

ONTOLOGICAL ODDITIES
Objects physically existing in one reality, but conceptually belonging to another reality

system, possible world, or real–imagined, fictional, or otherwise. Usually contradictory, sometimes paradoxical, and always strange. The fruits of extreme world-building.

ONTOLOGICAL ROLE-PLAY
Allows the designer to imagine their way into different "worlds" they can design from. Worlds that ideally do not quite make sense according to prevailing conceptual frameworks. These designs are not meant to be implemented. Their purpose is to open up new imaginative vistas onto the not here, not now through speculative ontologies materialized as everyday artifacts made strange.

ONTOLOGICAL SELF-DETERMINATION
In relation to Indigenous peoples, accepting that they do not need to articulate their concepts or ontologies in Western terms.[30]

ONTOLOGICAL SHOCK
A term used by Bernardo Kastrup in *Meaning in Absurdity* in a section discussing speculative ontology and the work of Jacques Vallée, who defined it as "the mechanism by which the phenomenon forces an expansion of people's conception of reality towards a worldview where notions previously held to be absurd become intelligible."[31] A deeper kind of estrangement.

ONTOLOGICAL TURN
Centers on the question of whether different "worlds" are merely different perspectives or views on one world typically built from Western ideas (one world, many worldviews), or whether worlds in fact constitute different ontologies. A helpful but underused concept for building worlds to thinking with and through rather than being implemented. Gets interesting when different ontologies collide with the material world, infusing objects with a fundamentally different status. Objects are physically present in all worlds, but their reason for being and conceptual purpose belongs only to one.[32]

ONTOLOGIST
A character in Charles L. Harness's short story "The New Reality" charged with ensuring reality remains stable by suppressing any research that might lead to devices that somehow transcend the human umwelt and reveal the "thing in itself," or the "really real," leading to madness.[33]

ONTOLOGY
What can and cannot exist within a particular reality system, or real.

OTHERWISE
An alternative to "futures" as a way of framing the "not here, not now." Looser, more open, and free from ties to the here and now.

OUTSIDER ART

When talking about other words, ontological pluralism, parallel realities, and the not here, not now, it is hard not to think of outsider art. Strictly speaking, this is art produced by people experiencing some kind of mental illness, which in itself is a problematic view. Maybe they occupy unique and personalized worlds at odds with consensus reality. But when they produce art, it is often stunning in its originality and imaginative richness. The most interesting examples use images and objects that hint at almost unimaginable realities or ways of making sense of the world.

OVERTON WINDOW

What is currently regarded as acceptable by the general public in relation to policymaking. Can be shifted in different directions so the “unthinkable” becomes the everyday. Can happen by accident too.

PANPSYCHISM

The belief that matter has an extremely basic form of consciousness. Not in the same sense as humans do, but in some other, as of yet unknown form. For some, this is a logical and possibly only conclusion for those who reject dualism, and embrace materialism as the home of consciousness, the self, “I,” or “me.”

PARADIGM SHIFT

A term coined by physicist and philosopher Thomas Kuhn to describe a significant and sudden shift in the understanding of how reality works, especially from a scientific point of view. For example, Nicolaus Copernicus’s discovery that the earth is not at the center of the universe and in fact rotates around the sun, or Albert Einstein’s theory of relativity. Interestingly, the world out there is just as it always was; it is the description of it and its workings that changes. But from a human perspective, the world changes too, and how we relate to it from then on.

PARADOX

Can a physical object be paradoxical, or is it a quality found only in things made from words and images?

PARALLEL WORLD/UNIVERSE

Grounding a narrative in the familiarity of the here and now by taking a step sideways rather than into the future or past. Also allows more latitude for adjustment to the laws of nature, such as than a counterfactual or alternative history might, which would require a stronger link to existing reality.

PATAPHYSICS

Invented by Alfred Jarry in the early twentieth century, and often defined as the science of imaginary solutions. It rarely surfaces in design. Is using it as a label a contradiction?

PERSPECTIVISM
Different views onto one (shared) world. The issue being, Whose world? In contrast to the ontological turn in anthropology, which advocates for the existence of many worlds rather than many perspectives onto one world.

PLURIONTOLOGY
One of Yuk Hui's variations on the idea of a pluriverse. Interesting for us because it suggests a role for design in making multiple ontologies tangible in order to prompt new thinking and perspectives. Remains slightly abstract and free from the political realities of how related concepts such as the pluriverse might play out in the actual world.

PLURIVERSAL DESIGN
Ontological plurality *within* design and the value this brings to wider discussions about the kinds of world(s) people wish to live in. Pluriversal design as opposed to designing for the pluriverse.

PLURIVERSE
A term with many origins and many contexts. Design usually references its use by anthropologist Arturo Escobar to describe a situation where different, usually overlooked cultural and social realities can coexist, supported by a form of autonomous design free from consumerist and modernist agendas. A pushback against the idea that we all live in one universal, human-centric world defined by Western concepts.

POCKET UNIVERSE
A term coined by theoretical physicist and cosmologist Alan Guth and associated with string theory. Rather than one universe, there could be many. Each with its own laws of physics and even different numbers of dimensions, which could be ten, eleven, or twenty-six according to various theories. Also, a region or pocket of space over which the laws of physics are uniform. In speculative fiction, pocket universe is used to describe a universe contained within another one with its own laws of physics. Pocket dimension, a variation, is used to describe an auxiliary dimension where a small space might contain a larger one. Also arises in discussions around the possibility or impossibility of creating a self-contained universe in a lab, sometimes called cosmogenesis or universe making. But what would it look like? Some say it would take the form of a tiny wormhole and that we would see only its "mouth."

POLYREALISM
Another concept from Yuk Hui. Again, usefully sidesteps the political challenges of a pluriverse, although it can easily suggest a world where fake news,

post-truth and even one's own reality dominate.

POSSIBLE
"Is it possible?" "That's not possible." "Let's make it possible." Always at the heart of a certain kind of design conversation. One that from the start, expects design to be aligned with existing reality, the here and now. A red herring for those of us concerned with catalyzing new ideas, thinking, and practice.

POSSIBILIA
Philosopher David Lewis's term for people and other entities inhabiting possible worlds.

POSSIBILISM
According to possibilism, fictional objects (or possibilia) exist even if not in the actual world. They "are," so they exist somewhere just as people and chairs do. But that somewhere is an infinite collection of "possible worlds" completely detached from and discontinuous with our own, both in terms of time and space. So where do they exist? In comparison, the way possible worlds are handled in design seems overly cautious and conservative.

POSTNORMAL TIMES
Coined by Ziauddin Sardar to describe the current in-between situation many people are currently experiencing, especially in the West. A "period where old orthodoxies are dying, new ones have not yet emerged, and nothing really makes sense."[34]

PRETERNATURAL
Beyond what is normal or natural (dictionary definition). But still an actual phenomena in that it can be experienced or encountered, although it cannot be explained by science. Appears to lie beyond the laws of nature. Strange and unexplainable. Preternatural beauty or silence, for example.

PROBABILITY WAVE
A quantum mechanical tool for working with phenomena at the tiny scale. Does not predict where an electron will be, for example, but where it is likely to appear if observed correctly. That is, if you look for it; otherwise it remains fuzzy.

PROPAGANDA
An amplification and weaponization of an already existing, usually negative aspect of the here and now. Rarely completely made up or invented. Possibly obsolete in the age of social media and post-truth, where contemporary forms of propaganda no longer merely attempt to influence reality or offer counternarratives but instead to shred the fabric of reality itself.

PROPOSITION
Probably what a design proposal should actually be called. Offering something for consideration.

A little looser than a proposal due to its conjectural or hypothetical nature.

PSEUDO-ENVIRONMENT
Journalist Walter Lippmann's term for the simplified model of a world people carry around inside them. Made from stereotypes, myths, and other generalizations gleaned from popular media, and is, as expected, riddled with bias and prejudice. Yet it is used to navigate and make sense of the complex political and social reality people find themselves inhabiting.

QUANTUM COMPUTER
An assemblage of the full gamut of Meinongian objects. Currently a semi-impossible object located on the border of reality, here, now, but also elsewhere. Stops working as soon as "reality" impinges on it. Great lengths must be taken to isolate it from its environment.

QUANTUM INTERPRETATION
There is much debate over whether quantum mechanics describes actual reality, revealing it to be far more strange than we ever thought, or if it is a toolbox that allows incredibly accurate operations to be carried out on reality, giving rise to so many of the technologies we depend on today. In response to this, there are many so-called interpretations of how quantum mechanics works. For example, the Copenhagen interpretation, De Broglie-Bohm theory, quantum information theories, QBism, objective collapse theories, and so on. Despite over a hundred years of debate, no consensus has been reached, although the Copenhagen interpretation is the most widely accepted. The most controversial is Hugh Everett's MWI.

QUANTUM MECHANICS
A theory that possibly describes reality and how it works, but definitely provides a toolbox for physicists to work with the very small, incredibly accurately.

QUANTUM MYSTICISM
Considered quackery by most physicists. Different from quantum cosmology, which attempts to understand the universe through a quantum lens.

"REAL"
Shorthand for a reality system. Its awkwardness forces a reader to pause for a moment and ask, not what is real, but what is a "real." It feels closer to the idea of a "world" than a "reality," and to us at least, suggests something belonging more to the realm of shared inner worlds than immutable physical worlds located outside the human mind.

REALISM
A Victorian idea concerned with writing fiction in a style that reflects existing reality. Dominates design, even when fictional.

REALIST
Someone who accepts the world as it is.

REALITY
To put it mildly, reality is not something that can be easily defined, even by philosophers and physicists, who spend much of their time thinking about its nature. The context we are expected to design for or within. The here and now. When we use "reality" in this book, we mean the complex fusion of ideas, stories, assumptions, and materiality, both actual and imagined, that constitute the shared "world" a person feels part of mentally, socially, culturally, and physically. Different from whatever lies beyond our human sensory world.

REALITY ENFORCEMENT
The forcing on people of a consensus reality. A tendency more present in design than fine art and maybe one of the biggest differences between the fields. Designers are expected to align their work and thinking with existing reality, and in some respects reinforce it. Artists are not only free to ignore it but also are often expected to challenge it, or reveal its joints and seams.

REALLY REAL
Is anything really real? Or maybe a realm lying beyond the human sensorium we do not have access to.[35]

REALITY SYSTEM
Something human-made rather than found, overlaid on whatever is out there. Can be unmade and remade. There have been many, and hopefully new ones will continue to evolve.

"REAL WORLD"
In design, often used in false opposition to places where imaginative work happens, such as academia, the art world, cultural organizations, and so on. Assumes that there is an inside and outside to reality. Imaginative work is somehow located outside reality in a separate realm, if that is possible.

SCI-PHI
Science fiction as philosophy. See, for example, the *Sci Phi Journal*, which describes it as "idea-driven fiction, as opposed to the 'character-driven' mode that has come to predominate speculative fiction."[36] Variations of which are sometimes glimpsed in the darker corners of design fiction.

SIMULATION
In an age of deep fakes, fake news, post-truth, and other related phenomena, a relatively straightforward and uncontroversial artificial construct located in the here and now. And by today's standards, easily understood as such.

SPACE-TIME
A setting for our lives and actions, in the West at least. Possibly a product of the human umwelt or maybe the material the human umwelt is made from.

SPECULATION
There are many reasons to speculate in design, good and bad: to mitigate risk by generating pathways through uncertain futures; presenting possible solutions (speculative solutionism) to a problem; developing alternative futures from which a community can choose; visualizing or materializing a shared goal; exploring a technology's social, political, or ethical implications; public relations and self-promotion; visualizing both propaganda and counternarratives; cautionary tales; expanding the public capacity for imagination; enlarging shared realities by introducing new and previously unthought possibilities; acclimatizing to technologies before they arrive; providing complicated pleasures; entertaining; escaping; nourishing inner worlds; and stimulating thought. Each has its own history, trajectory, context, and methods. For us, it should always be troublesome, challenging, and gently provocative.

SPECULATIVE FABULATION
According to scholar Donna Haraway, a "mode of attention, a theory of history and a practice of worlding."[37]

SPECULATIVE PRACTICES
Practices in any field that step outside the bounds, norms, and givens of existing reality, or the here and now, to explore alternative possibilities and impossibilities, imagined or actual.

SPECULATIVE REALISM
Turning to Levi R. Bryant again, "Speculative realism is a loosely affiliated philosophical movement that arose out of a Goldsmith's College conference organized by Alberto Toscano in 2007. While the participants at this event–Ray Brassier, Iain Hamilton Grant, Graham Harman, and Quentin Meillassoux–share vastly different philosophical positions, they are all united in defending a variant of realism and in rejecting anti-realism or what they call 'correlationism.'"[38]

SPOOKY ACTION AT A DISTANCE
Albert Einstein's name for quantum entanglement. An intriguing example is offered by Philip Ball in his book *Beyond Weird*: "The main thing you need to know about entanglement is this: it tells us that a quantum object may have properties that are not entirely located on that object."[39] The basis of a still theoretical quantum internet currently being developed in China, which holds the current record for spooky action

at a distance (1,203 kilometers between Delingha and Lijiang).

SUBJECTIVE IDEALISM
A philosophy developed by George Berkeley that claims nothing can exist independently of the mind. Sometimes known as empirical idealism or immaterialism. For an example of how this might manifest in literary world-building, see Jorge Luis Borges's "Tlön, Uqbar, Orbis Tertius."

SUBSIST
From Alexius Meinong's taxonomy of objects. Objects that exist, but not in space and time. Real but nonactual. Based on his idea that everything in the universe, including things that can only be thought such as unicorns and round squares, exist, just in different ways. To subsist means to be sort of real–that is, it exists, but is nonactual.

SUPERNATURAL
Unexplainable by the laws of nature or science (dictionary definition). Describes something that does not exist or belong in the natural world. The realm of ghosts, gods, spirits, and so on. A not here, not now to be handled with care.

SUPERPOSITION
When quantum objects are in multiple states. For example, a classical bit is either zero or one, whereas a qubit–that is, a quantum bit–can be zero, one, and everything in between at once. Also describes a state where entities can exist in two places at once. Currently limited to the scale of quantum objects such as photons.

SURREALISM
When inner worlds find form in the external world, undiluted. Entertaining, weird, but not always interesting. More useful as a diagnosis of the state of the collective imagination expressed through images, objects, and sometimes design.

SUSPENSION OF DISBELIEF
A deal between the fictioneer and consumer of fictions. Sometimes achieved involuntarily, but always more interesting when done by choice.

THEORY-OBJECT
An object made from a fusion of matter and ideas, closer to a model than a prototype. Like other objects focused on materials, the content is primarily about its materiality and how it is crafted–in this case, thoughts made from values, belief systems, ideas, and concepts. Its main purpose is to provide "complicated pleasure." Designed to get people thinking. They sit in the world slightly differently from other designed objects and serve a different purpose, requiring different forms of interaction and engagement when encountered.

THING

A looser way of thinking about objects that acknowledges their existence as material nodes for entangled systems available to the human senses.

THOUGHT EXPERIMENT

Most interesting when made from words. A useful fiction designed to test an idea that cannot be tested in the here and now.

TRUTH

An unhelpful term in design, often used to mean “the real” or “actual reality.”

UCHRONIA

A temporal fiction, an unspecified time period distant from that of the reader. Similar to utopia, but an else-when rather than an else-where. Not as precise as a counterfactual as it is not necessarily set in a recognizable time frame, either past or future.

UMWELT

As humans, we assume there is only one world, and that all animals share that world. But we can only ever experience an edited version of whatever is out there–what we think of as reality–filtered through our senses. Not just humans, but each life-form experiences its own unique world based on its senses and cognitive apparatus. In his 1934 essay “A Stroll through the Worlds of Animals and Men: A Picture Book of Invisible Worlds,” Jakob von Uexküll called this sensory world an umwelt. Rather than thinking about how to invite nonhumans into a human world, maybe first we need to imagine how we appear or are present in nonhuman worlds. How we are transformed, materially and conceptually, if we acknowledge nonhuman worlds based on different senses–olfactory, electric, seismic, magnetic, or auditory.

UNCANNY

A term used to describe situations that we rarely, if ever, experience. Estrangement plus something more. Not just strange, but uncanny–reality flexes for a moment, appearing to do something it shouldn’t. Encountering an apparent doppelgänger. Extreme coincidence. Strangeness that doesn’t quite make sense, then does, but is still strange, leaving a lingering feeling–maybe this is uncanniness.

UNIMAGINABLE

The impossibility of creating a picture of something in the mind. For many people, things being otherwise is unimaginable. Can design help to make the unimaginable imaginable?

UNKNOWABLE

As Norwegian novelist Karl Ove Knausgaard suggests in *Morning Star*: “It is never the case that we know what we see, but rather the other way round: we see what we know.”[40] How can we know about what is not yet known? Is there a limit to human intelligence where eventually it will just not be possible to know more than the mind allows? Might we reach a moment in time when the unknowable solidifies into a forever inaccessible realm? Or are there continuously evolving intellectual capacities and techniques that will always push the boundaries of what is knowable? Science of course being one such technique. Philosophy another.

UNPICTURABLE OBJECTS

A category of objects mentioned by art historian James Elkins in *Six Stories from the End of Representation* in relation to quantum mechanics where he draws a distinction between something being "merely unrepresentable" and being entirely "unavailable to the imagination as pictures."[41]

UNREAL

In design, a term used to dismiss anything that does not reinforce the existing order of things by embracing realism and aligning with how things are now. For us, an important idea that needs to be explored more. Not as escape or rejection of reality, but as another kind of thinking space in which to develop ideas that make reality a little larger.

UNTHINKABLE

A difficult concept to discuss, when you really think about it. Surely once something enters our mind it becomes a thought and is no longer unthinkable. A tantalizing concept suggesting a world of ideas lying just over the horizon of what can currently be thought. Are there yet to be discovered forms of thought that can reveal the currently unthinkable?

UTOPIA

A problematic concept giving rise to much confusion, at least in design. The title of a work of fiction by Thomas More written in 1516. Often considered an ideal to be aspired to. But whose ideal, and what about those for whom it is not ideal? Historically, many utopias led to actual dystopias. Some would say our current situation in the West is a kind of failed techno-utopia that began at the end of the last century. A more helpful definition is utopia as an impossible place or no place. For us, utopias are not meant to be realized; it is their impossibility that makes them useful.

VEGETABLE LAMB

For many years rumors of a zoophyte, said to exist in Central Asia, circulated throughout the West. See, for example, Henry Lee's *The Vegetable Lamb of Tartary* (1887). A fictional addition to Ziauddin Sardar's "Menagerie of Postnormal Potentialities" that includes black swans, black elephants, and black jellyfish, among others, and describes different tensions between the known, unknown, and unknowable.[42]

The vegetable lamb is a speculative and critical category for ideas that although seemingly reasonable today, might retrospectively be shown to be ridiculous.

VERISIMILITUDE

Staying true to (prevailing) reality in fictional representations. Often aspired to in design fiction, but leading to either hoaxes or too easy assimilation of the not here, not now into the here and now.

VITAL MATERIALISM

Also "vital materiality" and "vibrant matter." A contemporary view grounded in new materialism where the agency of objects and their active participation in the shaping of the human world can be explained through a form of "vitality" or "aliveness." Possibly real, possibly metaphoric.

WEIRD

The best definition we have found is by Mark Fisher in The Weird and the Eerie: "[T]he weird is a particular kind of perturbation. It involves a sensation of wrongness: a weird entity or object is so strange that it makes us feel that it should not exist, or at least it should not exist here. Yet if the entity or object is here, then the categories which we have up until now used to make sense of the world cannot be valid. The weird thing is not wrong, after all: it is our *conceptions* that must be inadequate."[43]

WORLD

The world is the set that contains everything. But is that really possible? Surely there is a something that lies outside the world? If not, then as philosopher Marcus Gabriel argues, perhaps the world does not exist, cannot exist, conceptually that is. Yet like Gabriel, for us, the world contains not just actual things but also ideas, numbers, and fictional entities, which all exist somewhere, even if it is in the imagination, in our minds, which is part of the world too. Or "a world is, so to say, the 'limit' of a series of increasingly more inclusive situations."[44] For us, a "world" encapsulates a set of beliefs, ideas, hopes, and fears that provide a framework for making sense of things, of how things work conceptually as well as what is and is not possible. A sort of useful fiction linking individual minds to a larger collective imagination that shapes the "contours of the possible."[45]

WORLD-BUILDING

The design and construction of the world in which a story or film is set. Can include alternative laws of physics, cosmology, culture, language, justice, economics, biology, and so on. Usually consists of familiar elements assembled in new ways.

WORLDING

Although the term is associated with Gayatri Spivak and decolonial studies, it is artist Ian Cheng's use of worlding that we are interested in here. It deals with seeing and making worlds. Of empowering people to recognize a multitude of worlds, create their own world(s), and participate in a variety of worlds. It is about being literate and having agency through an acknowledgment that we are living in a time of many worlds. A democratization of the not here, not now.

WORLD MAKING
Originally coined by philosopher Nelson Goodman in 1978. A form of meaning making where people make their own worlds. More of a literary form of world-building that happens in the mind of the reader than in the words of the writer. Can be triggered by words, music, dance, or visual representations. Helpful in shifting the way viewers approach designs for the not here, not now.

WORLDVIEW
What exactly is a worldview? It feels impossible to define precisely. Made from a fusion of ideas, assumptions, values, and beliefs, it is more expansive than a mindset, more functional than an imaginary, more fuzzy than an ideology, and more compact than a cosmology.

9 C/D: By a Conclus

Way of
ion

Chapter 9

C/D: By Way of a Conclusion

We began our book *Speculative Everything* with a list we called A/B, where “A” characterized how most people thought of design at that time, and “B” suggested a set of contrasting values. It was never intended to be fixed, and in fact we suggested that it could be followed by C, D, E, and so on, and could evolve.

So much has changed since then. In ourselves, design, and the world. We have decided to end this book with a similar list. Part description of the reality we find ourselves designing, teaching, and thinking for, and partly a suggestion for new ways of thinking about and practicing design based on ideas we have explored in this book.

C	D
Here, Now	Not Here, Not Now
Dogma	Doubt
New Normals	Post Normal[1]
Extrapolation	Parallelism
Known Unknowns	Unknown Unknowns
Science Fiction	Philosophy Fiction
Real World	A Larger Reality
Prototype	Model
Anthropocentric	More than Human
Newtonian Mechanics	Quantum Mechanics
What if	As if
Alternative Futures	Speculative Ontologies
Design for the Pluriverse	Pluriversal Design
Utopia / Dystopia	Heterotopia
Either/Or	And/Both
Prescription	Prompt
Design Theory	Theoretical Design
Technological Reason	Cosmotechnics[2]
Possible Objects	Impossible Objects

Preface

1. A phrase we borrowed from Susan Strehle, *Fiction in the Quantum Universe* (Chapel Hill: University of North Carolina Press, 2000).

Chapter 1: Introduction: Realists of a Larger Reality

1. Timothy Morton, "Beginning after the End," in *Dark Ecology: For a Logic of Future Coexistence* (New York: Columbia University Press, 2016), 1-2. Quoted in Dan Byrne-Smith, ed., *Science Fiction* (Cambridge, MA: MIT Press, 2020), 200.

2. For a thoughtful reflection on other uses of fiction in design, see Björn Franke, "Design Fiction Is Not Necessarily about the Future" (paper presented at the Swiss Design Network Conference: Negotiating Futures, Design Fiction, Basel, Switzerland, October 28-30, -2010), https://www.bjornfranke.com/_texts/franke_2010_design_fiction_is_not_necessarily_about_the_future.pdf. And for an expansive survey of different approaches to design as a form of material discourse, see Bruce M. Tharp and Stephanie M. Tharp, *Discursive Design: Critical, Speculative, and Alternative Things* (Cambridge, MA: MIT Press, 2019).

3. Mark Fisher, *Capitalist Realism: Is There No Alternative?* (Winchester, UK: Zero Books, 2009).

4. The referendum result was announced on Friday, June 24, 2016, at 07:20 BST, and the United Kingdom eventually withdrew from the European Union at 23:00 GMT on January 31, 2020.

5. For an excellent interview with Miéville about *The City and the City* that touches on conceptual parallels to real-world situations such as Baarle-Nassau and Baarle-Hertog, see Geoff Manaugh, "Unsolving the City: An Interview with China Miéville," BLDGBLOG (blog), March 1, 2011, https://bldgblog.com/tag/china-mieville/.

6. There are of course many different ways of making sense of the world besides Western systems of thought that we just do not have access to on a deep enough level. While acknowledging their existence and importance, we do not want to embark on a misguided form of metaphysical appropriation. It is our hope, though, that the underlying ideas in this book can be used to open up new possibilities for design in many different cultural contexts besides those, we the authors, are working within.

7. Although several design theorists have explored the idea of ontology in relation to design, there are still few, if any, examples of what this might look like when translated into design practice. For a helpful overview of how design theory has engaged with ontology, see J. P. Hartnett, "Ontological Design Has Become Influential in Design Academia–But What Is It?," *AIGA Eye on Design*, June 14, 2021, https://eyeondesign.aiga.org/ontological-design-is-popular-in-design-academia-but-what-is-it.

8. In 2015, we made a video to accompany an edited lecture from carneades.org (with permission) about Meinong's ideas called *MAK Design Salon 4: Meinong's Taxonomy of Objects*. It can be viewed here: https://vimeo.com/133160620.

9. The project was accepted into the collection, but on the basis that the books they are described in are physical objects.

10. Hélène Schumacher, "Is This the Most Powerful Word in the English Language?," *BBC Culture*, December 31, 2020, https://www.bbc.com/culture/article/20200109-is-this-the-most-powerful-word-in-the-english-language.

11. Mark Fisher, *The Weird and the Eerie* (London: Repeater, 2017).

12. George Saunders, *A Swim in a Pond in the Rain: In Which Four Russians Give a Master Class on Writing, Reading, and Life* (New York: Random House, 2021), 383, Kindle.

13. Ronald Barnett, *Imagining the University* (Abingdon, UK: Routledge, 2013), 23, Kindle.

14. Ursula K. Le Guin, *No Time to Spare: Thinking about What Matters* (New York: Harper, 2017).

15. Reality is not something that can be easily defined, even by philosophers and physicists who spend much of their time thinking about its nature. When we use "reality" in this book, we mean the complex fusion of ideas, stories, assumptions, and materiality, both actual and imagined, that constitute the shared "world" a person feels part of, mentally, socially, culturally, and physically.

16. Barnett, *Imagining the University*, 34.

17. For some recent examples, see Silvio Lorusso, *What Design Can't Do: Essays on Design and Disillusion* (Eindhoven, Netherlands: Set Margins', 2024); Mike Monteiro, *Ruined by Design: How Designers Destroyed the World, and What We Can Do to Fix It* (San Francisco: Mule Books, 2019).

18. William Morris, *News from Nowhere* (1890; repr., Garden City, NY: Dover Publications, 2011), Kindle.

19. Ursula K. Le Guin, "Acceptance Speech for the National Book Foundation's Medal for Distinguished Contribution to American Letters" (filmed at the sixty-fifth National Book Awards, New York, November 2014), https://www.youtube.com/watch?t=12&v=Et9Nf-rsALk.

Chapter 2: An Archive of Impossible Objects

1. Robert E. Post, *American Enterprise: Nineteenth-Century Patent Models: An Exhibition Organized by Cooper-Hewitt Museum, the Smithsonian Institution's National Museum* (New York: Cooper-Hewitt Museum, 1984).

2. We first encountered the slightly awkward term "reals" in a paper by John Law called "What's Wrong with a One-World World?" Its awkwardness forces a reader to pause for a moment and ask not, What is real? but rather, What is *a* real? It feels closer to the idea of a "world" than a "reality," and to us at least, suggests something belonging more to the realm of shared inner worlds than immutable physical worlds located outside the human mind. The paper can be found on Heterogeneities, September 19, 2011,

http://www.heterogeneities.net/publications/Law2011WhatsWrongWithAOneWorldWorld.pdf.

3. Michel Foucault, *The Order of Things* (New York: Pantheon, 1970), xv.

4. Foucault, *The Order of Things*, xv.

5. Quoted in Hua Hsu, "How Sun Ra Taught Us to Believe in the Impossible," *New Yorker*, June 28, 2021, https://www.newyorker.com/magazine/2021/07/05/how-sun-ra-taught-us-to-believe-in-the-impossible.

6. See, for example, Lorraine Daston, ed., *Biographies of Scientific Objects* (Chicago: University of Chicago Press, 2000).

7. Darren Oldridge, *Strange Histories: The Trials of the Pig, the Walking Dead, and Other Matters of Fact from the Medieval and Renaissance Worlds* (Abingdon, UK: Routledge, 2004), 39.

8. Thomas G. Pavel, "Fiction and the Ontological Landscape," *Studies in 20th Century Literature* 6, no. 1 (1981): article 8, https://doi.org/10.4148/2334-4415.1630.

9. Matt Ward, "Design, Fiction and the Logic of the Impossible," *Medium*, July 18, 2019, https://medium.com/@matthewward/design-fiction-and-the-logic-of-the-impossible-7d10bfcf2b4b.

10. For more on possible worlds, see David K. Lewis, *On the Plurality of Worlds* (London: Wiley-Blackwell, 2001).

11. A video of the illusion can be viewed at https://www.youtube.com/watch?v=oWfFco7K9v8.

12. For more on this, see René Descartes, *Meditations on First Philosophy*, trans. J. Veitch (Toronto: Our Open Media, 2017), 51-52, quoted in Kristupas Sabolius, "Reality, Determination, Imagination," *Open Philosophy*, accessed February 21, 2024, https://doi.org/10.1515/opphil-2020-0122.

13. Tristan Garcia, *Form and Object: A Treatise on Things* (Edinburgh: Edinburgh University Press), 64, Kindle.

14. Graham Priest, "Beyond True and False," *Aeon*, May 5, 2014, https://aeon.co/essays/the-logic-of-buddhist-philosophy-goes-beyond-simple-truth. Priest offers an excellent, nonspecialist discussion of these and related ideas.

15. Vilém Flusser, "On Fiction," trans. José Newman, in *EP Volume 2: Design Fiction, ed. Alex Coles* (Berlin: Sternberg Press, 2016), 136-138.

16. Michel Foucault, "Of Other Spaces, Heterotopias," translated from Architecture, *Mouvement, Continuité 5* (1984): 46-49, https://foucault.info/documents/heterotopia/foucault.heteroTopia.en.

17. See Timothy Morton, *Hyperobjects: Philosophy and Ecology after the End of the World* (Minneapolis: University of Minnesota Press, 2013).

18. See Lorraine Daston, ed., *Things That Talk: Object Lessons from Art and Science* (Princeton, NJ: Princeton University Press, 2007).

19. See Bruno Latour, *Reassembling the Social: An Introduction to Actor-Network-Theory* (Oxford: Oxford University Press, 2007).

20. See Bill Brown, *A Sense of Things: The Object Matter of American Literature* (Chicago: University of Chicago Press, 2003).

21. See Karen Barad, *Meeting the Universe Halfway: Quantum Physics and the Entanglement of Matter and Meaning* (Durham, NC: Duke University Press, 2007).

22. See Graham Harman, *Object-Oriented Ontology: A New Theory of Everything* (London: Pelican Books, 2018), Kindle.

23. See Jane Bennett, *Vibrant Matter: A Political Ecology of Things* (Durham, NC: Duke University Press, 2010).

24. See Donald David Hoffman, *The Case against Reality: Why Evolution Hid the Truth from Our Eyes* (New York: W. W. Norton, 2019).

25. See Diana Coole and Samantha Frost, eds., *New Materialisms: Ontology, Agency and Politics* (Durham, NC: Duke University Press, 2010).

26. See, for example, Philip Ball, *Beyond Weird: Why Everything You Thought You Knew about Quantum Physics Is Different* (Chicago: University of Chicago Press, 2018), Kindle.

27. See, for example, Philip Goff, "The Case for Panpsychism," *Philosophy Now*, accessed March 31, 2023, https://philosophynow.org/issues/121/The_Case_For_Panpsychism.

28. For an overview of this approach in anthropology, see Paolo Heywood, "The Ontological Turn," Open Encyclopedia of Anthropology, May 19, 2017, http://doi.org/10.29164/17ontology.

29. Graham Priest, "A Short Story and Ten Morals," *Notre Dame Journal of Formal Logic* 38, no. 4 (Fall 1997): 573-582.

30. For a detailed discussion of this story within a philosophical context, see Daniel Nolan, "A Consistent Reading of Sylvan's Box," *Philosophical Quarterly* 57, no. 229 (October 2007): 667-673, https://www.jstor.org/stable/4543271.

31. Haruki Murakami, *First Person Singular: Stories*, trans. Philip Gabriel (New York: Vintage, 2021), 18, Kindle.

32. For an extensive list of artifacts and phenomena that appear in the story, see http://pid.bungie.org Roadside_Picnic_Artifacts.html.

33. Flan O'Brien, *The Third Policeman* (Funks Grove, IL: Dalkey Archive Press, 2001).

34. Joel Levy, *The Infinite Tortoise: The Curious Thought Experiments of History's Great Thinkers* (London. Michael O'Mara, 2016), 166, Kindle.

35. See, for example, Adrian Hon, *A History of the Future in 100 Objects* (New York: Skyscraper Publications, 2013).

36. "Impossible Object," Wikipedia, last modified November 27, 2022, https://en.wikipedia.org/w/index.php?title=Impossible_object&action=history.

37. See Chris Mortensen, *The Impossible Arises: Oscar Reutersvärd and His Contemporaries* (Bloomington: Indiana University Press, 2022).

38. Philipp used 205 images that he manually downloaded, increasing the number to 615 by computationally modifying them slightly–for example, mirroring, cropping, and shifting. Those 615 were used to train the GAN, a common practice called data augmentation. It is a small sample, but it was enough for us to get a feeling for how this might work as an experiment.

39. From a conversation about the project with James Foster at Apple Design UK, November 24, 2021.

40. According to writer Benjamín Labatut, "Schrödinger too came to detest quantum mechanics. He contrived an elaborate thought experiment, a Gedanken-experiment, the result of which was an apparently impossible creature: a cat that was, at once, alive and dead." Benjamín Labatut, *When We Cease to Understand the World*, trans. Adrian West (New York: New York Review of Books), 170, Kindle.

41. Rivka Galchen, "Dream Machine: The Mind-Expanding World of Quantum Computing," *New Yorker*, May 2, 2011, https://www.newyorker.com/magazine/2011/05/02/dream-machine.

42. Victoria Gill, "'Whitest Ever' Paint Reflects 98% of Sunlight," *BBC News*, April 16, 2021, https://www.bbc.com/news/science-environment-56749105.

43. For a list of fictional colors, see https://list.fandom.com/wiki/List_of_fictional_colors.

44. Douglas Adams, *The Hitchhiker's Guide to the Galaxy* (New York: Del Rey Books, 1995), 68; David Toomey, *Weird Life: The Search for Life That Is Very, Very Different from Our Own* (New York: W. W. Norton, 2013), 166, Kindle.

45. Jorge Luis Borges, "Tlön, Uqbar, Orbis Tertius," in *Labyrinths*, trans. James Irby (New York: New Directions, 1962), 7.

46. The original quote reads, "The idea of universes with alternative laws of nature seems like the stuff of science fiction. But the truth is more mundane than it sounds. Modern medical technology routinely produces alternative universes inside MRI machines." Leonard Susskind, *The Cosmic Landscape: String Theory and the Illusion of Intelligent Design* (Boston: Little, Brown and Company, 2008), 91, Kindle.

47. See, for example, Zeeya Merali, "Creating a Universe in the Lab? The Idea Is No Joke," *Discover*, June 19, 2007, https://www.discovermagazine.com/the-sciences/creating-a-universe-in-the-lab-the-idea-is-no-joke.

48. Jonathan Lethem, *As She Climbed across the Table: A Novel* (New York: Vintage, 1998).

49. Denis Overbye, "Physicists Create 'the Smallest Crummiest Wormhole You Can Imagine,'" *New York Times*, November 30, 2022, https://www.nytimes.com/2022/11/30/science/physics-wormhole-quantum-computer.html.

50. John D. Barrow, *Impossibility: The Limits of Science and the Science of Limits* (Oxford: Oxford University Press, 1999), 72.

51. Adam Roberts, *The Thing Itself* (London: Gollancz, 2015), 100-101, Kindle.

52. For a detailed discussion of this, see Steven Shaviro, *Extreme Fabulations: Science Fictions for Life* (Cambridge, MA: MIT Press, 2021), 21-38, Kindle.

53. Roberts, *The Thing Itself*, 231-232.

54. Shaviro, *Extreme Fabulations*, 5.

55. Kyle Vanhemert, "Beautifully Warped Landscapes from Apple's Glitchy Maps App," *Wired*, July 1, 2013, https://www.wired.com/2013/07/beautifully-warped-landscapes-from-apples-glitch-prone-maps-app/; Peder Norrby's Flickr site, accessed February 9, 2023, https://www.flickr.com/photos/pedernorrby/sets/72157632277119513.

56. For more examples, see the work of artist Clement Valla, who collects Google Earth images gone wrong: http://www.postcards-from-google-earth.com. See also Clement Valla, "The Universal Texture," *Rhizome*, July 31, 2012, https://rhizome.org/editorial/2012/jul/31/universal-texture.

57. Barrow, *Impossibility*, 25.

58. Mark Twain, *Following the Equator: A Journey around the World* (1897; repr., Mineola, NY: Dover Publications, 1989).

59. José Saramago, *The Stone Raft*, trans. Giovanni Pontiero (New York: HarperVia, 1996).

60. Darko Suvin, "On the Poetics of the Science Fiction Genre," *College English* 34, no. 3 (December 1972): 365, http://www.jstor.org/stable/375141.

61. Darko Suvin, "Estrangement and Cognition," *Strange Horizons*, November 24, 2014, http://strangehorizons.com/non-fiction/articles/estrangement-and-cognition.

62. Darko Suvin, "Considering the Sense of 'Fantasy' or 'Fantastic Fiction': An Effusion," *Extrapolation* 41, no. 3 (October 2000): 209-247, https://doi.org/10.3828/extr.2000.41.3.209.

63. Jorge Luis Borges, "This Craft of Verse Lectures" (1967-68 Norton Lectures on Poetry, Harvard University, Cambridge, MA), 4:13:48, accessed June 28, 2023, https://www.youtube.com/watch?v=YSLV7t9DvN8.

64. Alan Lightman, *Einstein's Dreams* (1992; repr., New York: Vintage Books, 2011), Kindle.

65. For example, time passes at a different speed at the height satellites travel, which needs to be calculated using the theory of relativity.

66. George Gamow, *Mr Tompkins in Paperback* (Cambridge: Cambridge University Press, 1993), Kindle.

67. Robert Gilmore, *Alice in Quantumland: An Allegory of Quantum Physics* (Göttingen: Copernicus Publications, 1995).

68. For an in-depth exploration of designing for preexisting fictional literary worlds as a design method, see Austin T. J. Houldsworth, "For Money's Sake!: Introducing Redefinition Design–a Method to Break Out of the Ubiquitous Monetary Paradigm; in the Hope of Finding Genuine Alternatives" (PhD diss., Royal College of Art, September 2017), https://researchonline.rca.ac.uk/3537/1/Houdsworth PhD thesis(CopyRight Exclusions).pdf.

69. The project can be viewed at https://www.calendarcollective.com.

70. "Shukuchi," Wikipedia, last modified October 7, 2022, https://en.wikipedia.org/wiki/Chukjibeop.

71. "Moon Kyungwon and Jeon Joonho at the Korean Pavilion at the Venice Biennale," *e-flux*, April 14, 2015, https://www.e-flux.com/announcements/29700/moon-kyungwon-jeon-joonho-at-the-korean-pavilion-at-the-venice-biennale.

72. Frédéric Neyrat, "The Black Angel of History," trans. & trans. Daniel Ross, *Angelaki* 25, no. 4 (2020): 120-134, doi: 10.1080/0969725X.2020.1790841.

73. Quoted in Perrin M. Lathrop, "Telling Secrets: Abu Bakarr Mansaray's *Sinister Project*," *post*, MoMa, January 8, 2020, https://post.moma.org/telling-secrets-abu-bakarr-mansarays-sinister-project.

74. Yuk Hui, *The Question concerning Technology in China: An Essay in Cosmotechnics* (Falmouth, UK: Urbanomic, 2016), xiii, Kindle.

75. Yuk Hui and Pieter Lemmens, eds., *Cosmotechnics: For a Renewed Concept of Technology in the Anthropocene* (Abingdon, UK: Routledge, 2021), 119, Kindle.

76. For a detailed discussion of this technology, see Joseph C. Chapman and Nicholas A. Peters, "Paving the Way for Satellite Quantum Communications," *Physics*, November 9, 2022, https://physics.aps.org/articles/v15/172.

77. For more on this, see Karen Kwon, "China Reaches New Milestone in Space-Based Quantum Communications," *Scientific American*, June 25, 2020, https://www.scientificamerican.com/article/china-reaches-new-milestone-in-space-based-quantum-communications.

78. See Brian Hart, Bonny Lin, Samantha Lu, Hannah Price, Yujie (Grace) Liao, and Matthew Slade, "Is China a Leader in Quantum Technologies?," *China Power*, August 14, 2023, https://chinapower.csis.org/china-quantum-technology.

79. David M. Eagleman, "What Scientific Concept Would Improve Everybody's Cognitive Toolkit?: The Umwelt," *Edge*, 2011, https://www.edge.org/responses/what-scientific-concept-would-improve-everybodys-cognitive-toolkit.

80. Eric C. H. De Bruyn and Sven Lütticken, eds., *Futurity Report* (Cambridge, MA: MIT Press, 2020), 107; J.B.S. Haldane, *Possible Worlds and Other Essays* (London: Chatto and Windus, 1927).

81. Jakob von Uexküll, "A Stroll through the Worlds of Animals and Men: A Picture Book of Invisible Worlds," in *Instinctive Behavior: The Development of a Modern Concept*, trans. and ed. Claire H. Schiller (New York: International Universities Press, 1934), 9, https://monoskop.org/images/1/1d/Uexkuell_Jakob_von_A_Stroll_Through_the_Worlds_of_Animals_and_Men_A_Picture_Book_of_Invisible_Worlds.pdf.

82. Haldane, *Possible Worlds and Other Essays*, 268.

83. von Uexküll, "A Stroll through the Worlds of Animals and Men," 14.

84. Thomas Nagel, "What Is It Like to Be a Bat?," *Philosophical Review* 83, no. 4 (October 1974): 435-450, https://doi.org/10.2307/2183914.

85. "Historical Testament to Friendly Exchanges," Sohu, June 8, 2018, https://www.sohu.com/a/234676639_269136.

86. See, for example, Adam Fabio, "Theremin's Bug: How The Soviet Spied on the US Embassy for 7 Years," *Hackaday*, December 8, 2015, https://hackaday.com/2015/12/08/theremins-bug.

87. For a list of diplomatic gifts received on behalf of White House and Department of State officials by the Gift Office within the Office of the Chief of Protocol, see https://2009-2017.state.gov/s/cpr/c29447.htm.

88. For more on this, see Loraine Sievers, "Purposes, Politicisation and Pitfalls of Diplomatic Gift-Giving to the United Nations," *Hague Journal of Diplomacy*, February 4, 2021, https://brill.com/view/journals/hjd/16/1/article-p110_6.xml.

89. Diana Carrió-Invernizzi, "Gift and Diplomacy in Seventeenth-Century Spanish Italy," *Historical Journal* 51, no. 4 (December 2008): 883, https://www.jstor.org/stable/20175207.

90. Merlin Sheldrake, *Entangled Life: How Fungi Make Our Worlds, Change Our Minds, and Shape Our Futures* (London: Random House Trade Paperbacks, 2021), 23, Kindle.

91. Margaret W. Robinson, *Fictitious Beasts, a Bibliography* (London: Library Association, 1961).

92. Henry Lee, *The Vegetable Lamb of Tartary: A Curious Fable of the Cotton Plant; to Which Is Added a Sketch of the History of Cotton and the Cotton Trade* (1887; repr., Whitefish, MT: Kessinger Publishing, 2008).

93. Oldridge, *Strange Histories*, 167.

94. Bernardo Kastrup, *Meaning in Absurdity: What Bizarre Phenomena Can Tell Us about the Nature of Reality* (Winchester, UK: John Hunt Publishing, 2012), 74, Kindle.

95. Kastrup, *Meaning in Absurdity*, 75-76.

96. Claire Isabel Webb, "The Ladder, the Sphere and the Rhizome," *Noēma*, March 15, 2022, https://www.noemamag.com/the-ladder-the-sphere-and-the-rhizome.

Chapter 3: Quantum Common Sense: New Metaphors, Images, and Concepts?

1. See, for example, Alex Wilkins, "Quantum Computers Could Create Completely New Forms of Matter," *New Scientist*, June 9, 2022, https://www.newscientist.com/article/2323544-quantum-computers-could-create-completely-new-forms-of-matter.

2. See, for example, Marco Fellous-Asiani, "Could Energy Efficiency Be Quantum Computers' Greatest Strength Yet?," *Conversation*, October 25, 2023, https://theconversation.com/could-energy-efficiency-be-quantum-computers-greatest-strength-yet-191989.

3. For a nice explanation of the MWI, its history, and implications, especially for quantum computing, see John Gribben, "The Many-Worlds Theory, Explained," *MIT Press Reader*, May 20, 2022, https://thereader.mitpress.mit.edu/the-many-worlds-theory.

4. Rivka Galchen, "Dream Machine: The Mind-Expanding World of Quantum Computing," *New Yorker*, May 2, 2011, https://www.newyorker.com/magazine/2011/05/02/dream-machine.

5. Jennifer Rankin, "'Pathway' to Brexit Deal Hits Usual Obstacles on the Irish Border," *Guardian*, October 14, 2019.

6. Quoted in Quinn Norton, "The Father of Quantum Computing," *Wired*, February 15, 2007, https://www.wired.com/2007/02/the-father-of-quantum-computing.

7. "Carlo Rovelli: 'Time Travel Is Just What We Do Every Day...,'" *Guardian*, March 31, 2019.

8. Philip Ball, "Two Slits and One Hell of a Quantum Conundrum," *Nature* 560, no. 165 (August 7, 2018), https://www.nature.com/articles/d41586-018-05892-6.

9. Philip Ball, *Beyond Weird: Why Everything You Thought You Knew about Quantum Physics Is Different* (Chicago: University of Chicago Press, 2018), 19, Kindle.

10. Ted Chiang, *Exhalation: Stories* (New York: Vintage Press, 2019), Kindle.

11. Quoted in Matt Axvig, "Quantum Mechanics, Contingency, and Freedom in Ted Chiang's 'Anxiety Is the Dizziness of Freedom,'" *Veritas Journal*, August 11, 2019, https://veritasjournal.org/2019/08/11/quantum-mechanics-contingency-and-freedom-in-ted-chiangs-anxiety-is-the-dizziness-of-freedom.

12. More about the app can be found on the maker's website at https://www.aerfish.com/universe-splitter.

13. For more on this project, see https://gazell.io/exhibitions/95/works/artworks-8611-libby-heaney-cephalopod-alien-quantum-computing-study-2019-21.

14. For a detailed explanation of this, see Johnjoe McFadden and, Jim Al-Khalili, *Life on the Edge: The Coming of Age of Quantum Biology* (New York: Crown, 2015), 1-18, Kindle.

15. For a critical discussion of the use of quantum metaphors in different fields, especially literature, see Jennifer Burwell, *Quantum Language and the Migration of Scientific Concepts* (Cambridge, MA: MIT Press, 2018).

16. Michael P. A. Murphy, *Quantum Social Theory for Critical International Relations Theorists: Quantizing Critique* (London: Palgrave Macmillan, 2021), 3.

17. See their website at https://projectqsydney.com.

18. Alexander Wendt, *Quantum Mind and Social Science* (Cambridge: Cambridge University Press, 2015).

19. Fritjof Capra, *The Tao of Physics: An Exploration of the Parallels between Modern Physics and Eastern Mysticism* (Boulder, CO: Shambhala, 2010).

20. Karen Barad, *Meeting the Universe Halfway: Quantum Physics and the Entanglement of Matter and Meaning* (Durham, NC: Duke University Press, 2007), 24.

21. Nicholas Harrington, "Time Crystals and the Quantum Mindset," *Project Q* (blog), accessed March 23, 2023, https://projectqsydney.com/time-crystals-and-the-quantum-mindset.

22. Barad, *Meeting the Universe Halfway*, 17.

23. Ball, *Beyond Weird*, 163.

24. Albert Einstein, Boris Podolsky, and Nathan Rosen, "Can Quantum-Mechanical Description of Physical Reality Be Considered Complete?" *Physical Review*, 47, (1935), 777-780, http:/dx.doi.org/10.1103/PhysRev.47.777.

Chapter 4: Unreal by Design

1. Franco "Bifo" Berardi, *The Third Unconscious: The Psychosphere in the Viral Age* (London: Verso, 2021), 29.

2. Anthony Dunne and Fiona Raby, "The School of Constructed Realities," Maharam, accessed June 23, 2022, https://www.maharam.com/stories/raby_the-school-of-constructed-realities.

3. Joseph Voros, "The Futures Cone, Use and History," *Voroscope*, February 24, 2017, https://thevoroscope.com/2017/02/24/the-futures-cone-use-and-history.

4. Bernardo Kastrup, *Meaning in Absurdity: What Bizarre Phenomena Can Tell Us about the Nature of Reality* (Winchester, UK: John Hunt Publishing, 2012), 20-21, Kindle.

5. See, for example, Terry Winograd and Fernando Flores, *Understanding Computers and Cognition: A New Foundation for Design* (Boston: Addison-Wesley, 1987); Anne-Marie Willis, "Ontological Designing," *Design Philosophy Papers* 4, no. 2 (2006): 69-92, https://doi.org/10.2752/144871306X13966268131514; Arturo Escobar, *Designs for the Pluriverse: Radical Interdependence, Autonomy, and the Making of Worlds* (Durham, NC: Duke University Press, 2018); Joseph Lindley, Paul Coulton, and Haider Ali Akmal, "Turning Philosophy with a Speculative Lathe: Object-Oriented Ontology, Carpentry, and Design Fiction," in *Design as a Catalyst for Change: DRS International Conference 2018*, ed. Cristiano Storni, Keelin Leahy, Muireann McMahon, Peter Lloyd, and Erik Bohemia (Boston: Design Research Society, 2018); Ron Wakkary, *Things We Could Design: For More Than Human-Centered Worlds* (Cambridge, MA: MIT Press, 2021); Asvar Gurpinar, "Towards an Object-Oriented Design Ontology," in *DRS2022: Bilbao*, ed. Dan Lockton, Sara Lenzi, Paul Hekkert, Arlene Oak, Juan Sádaba, and Peter Lloyd (Boston: Design Research Society, 2022).

6. Steven Shaviro, endorsing John K. Shaw and Theo Reeves-Evison, eds., *Fiction as Method* (Cambridge, MA: MIT Press, 2017).

7. Ian Bogost, *Alien Phenomenology, or What It's Like to Be a Thing* (Minneapolis: University of Minnesota Press, 2012), 99, Kindle.

8. Quoted in Eric David, "For the Rest of Us: Hank Beyer

and Alex Sizemore Explore the Intangible Values of Regional Materials," *Yatzer*, June 9, 2020, https://www.yatzer.com/for-the-rest-of-us.

9. The project is beautifully documented in book form. See Hank Beyer and Alex Sizemore, *For the Rest of Us* (Montreal: Bookart, 2018).

10. "ADS4: Legal Fictions," Royal College of Art, accessed August 23, 2023, https://2021.rca.ac.uk/programmes/ads4-plots-props-paranoia--how-architecture-stages-conspiracy.

11. "ADS4: Legal Fictions–Dominic Oliver," Royal College of Art, accessed August 23, 2023, https://2021.rca.ac.uk/students/dominic-oliver.

12. See, for example, "Home Page," Ecovative, accessed April 27, 2023, https://www.ecovative.com.

13. For more about the project, see Daniela Fortino, "How the Mushroom Burial Suit Works and What Does It Cost," *Eirene*, October 11, 2022, https://eirene.ca/blog/how-mushroom-burial-suit-works.

14. Merlin Sheldrake, *Entangled Life: How Fungi Make Our Worlds, Change Our Minds, and Shape Our Futures* (London: Random House Trade Paperbacks, 2021), 215, Kindle.

15. For more information and the studio's projects, see http://www.departmentofseaweed.org.

16. Charles L. Harness, *The Rose* (New York: Berkley Medallion, 1969).

17. This section was first published in the catalog for the *NGV Triennial 23*, National Gallery of Victoria, Melbourne, Australia, and is an edited version of a talk given as part of the Prada Frames–Materials in Flux event in Milan, Italy, on April 18, 2023.

18. J. B. S. Haldane, *Possible Worlds and Other Essays* (London: Chatto and Windus, 1927), 265.

19. For favorite sources, see Larry Sultan and Mike Mandel, *Evidence* (1977; repr., New York: Distributed Art Publishers, 2017); Luce Lebart, *Inventions*, 1915-1938 (Paris: RVB Books, 2020).

20. Barbara Penner, Adrian Forty, Olivia Horsfall Turner, and Miranda Critchley, eds., *Extinct: A Compendium of Obsolete Objects* (London: Reaktion Books, 2021).

21. Other artists and designers have used archaeology in relation to fictional projects. See, for example, Elliat Rich, "Ode to Waratah: Designing Mythology," *Garland Magazine*, March 4, 2021, https://garlandmag.com/article/waratah; Julian Bleecker, *It's Time to Imagine Harder* (Los Angeles: Near Future Laboratory, 2023). We have also written about it ourselves in *Hertzian Tales*, but here, we are thinking about it as a generative rather than interpretative approach.

Chapter 5: A Public Lending Library of Things

1. Steven Shaviro, *The Universe of Things: On Speculative Realism* (Minneapolis: University of Minnesota Press, 2014), loc. 906, Kindle. "PR" is a reference to Whitehead's book *Process and Reality* (1929).

2. Shaviro, *The Universe of Things*, loc. 924, Kindle.

3. There are many other examples of conceptual libraries. For two favorites, see *The MacGuffin Library* (2008) by Onkar Kular and Noam Toran, http://www.onkarkular.com/index.php?/project/the-macguffin-library; *Interspecies Library* by Oscar Salguero, accessed March 26, 2024, https://interspecieslibrary.com.

4. Rebecca Solnit, *As Eve Said to the Serpent* (Athens: University of Georgia Press, May 2001), 164, cited in Michael Cherney, "Map of Mountains and Seas," Quimai.net, accessed September 28, 2022, https://www.qiumai.net/ecatalogs/map%20of%20mountains%20and%20seas.pdf.

5. Ursula K. Le Guin, *The Left Hand of Darkness: 50th Anniversary Edition* (London: Penguin Publishing Group, 2000), xi-xvi, Kindle; H. G. Wells, *The Scientific Romances of H. G. Wells* (London: Victor Gollancz, 1933); Quentin Meillassoux, *After Finitude: An Essay on the Necessity of Contingency*, trans. Ray Brassier (London: Continuum, 2010).

6. In the philosophy of science, see, for example, Ronald N. Giere, "How Models Are Used to Represent Physical Reality," *Philosophy of Science* 71, no. 5, *Proceedings of the 2002 Biennial Meeting of the Philosophy of Science Association, Part II: Symposia Papers* (December 2004): 742-752, https://doi.org/10.1086/425063.

7. Steffen Ducheyne, "Towards an Ontology of Scientific Models," *Metaphysica* 9, no. 1 (2008): 119-127, doi:10.1007/s12133-008-0026-y.

8. Surprisingly, little has been written about the model in design, especially as a medium. In his unpublished PhD dissertation, Björn Franke has a helpful section called "Thinking Things" that discusses models in some depth in relation to design as a form of philosophical inquiry. See Björn Franke, "Design as Inquiry: Prospects for a Material Philosophy" (PhD diss., Royal College of Art, 2016), 139-156. And despite a large body of work in fine art where models and sculpture overlap, it is difficult to find writing on this. It is probably in architecture, where models have served all kinds of purposes for centuries, that the most helpful sources on this topic can be found, which we draw on most often in this book. For a range of reflections on the model in architecture, see, for example, Cynthia Davidson, ed., *Log 50: Model Behavior* (New York: Anyone Corporation, 2020).

9. Christian Hubert, "The Ruins of Representation," in *Idea as Model: 22 Architects 1976/1980*, ed. Kenneth Frampton and Silvia Kolbowski (New York: Rizzoli, 1981), 17.

10. Teresa Fankhänel, *The Architectural Models of Theodore Conrad: The "Miniature Boom" of Mid-Century Modernism* (New York: Bloomsbury Visual Arts, 2021), 9.

11. For more on this project, see "Composite Presence–Biennale Architettura 2021," Flanders Architecture Institute, accessed March 7, 2023, https://www.vai.be/en/

expos-en-programma/composite-presence---biennale-architettura-2021.

12. Our partners on the project were Victoria Hattam (political science), Miriam Ticktin (anthropology), and T. Alexander Aleinikoff (immigration and refugee law). Between 2017 and 2019, the project brought together twenty faculty and students from across to university to imagine, discuss, and design new border forms. The work mentioned here, supported by A/D/O in Brooklyn, was completed during an additional phase focused on materializing ideas from the first part.

13. T. Alexander Aleinikoff, Anthony Dunne, Victoria Hattam, Fiona Raby, and Miriam Ticktin, "Imaginative Mobilities," proposal to the Mellon Foundation for a Sawyer Seminar Grant, March 31, 2017, 1.

14. From his study *The Moon: Considered as a Planet, a World, and a Satellite* (1874), for which he built models based on observations of the moon's surface made through a homemade telescope and then photographed them to show the most accurate representation of its topography at that time.

15. "The Cubical Planet: The Surprising Theory Based on Its Alleged Discovery," *New York Times*, November 16, 1884, https://www.nytimes.com/1884/11/16/archives/the-cubical-planet-the-surprising-theory-based-on-its-alleged.html.

16. For the original book, see Scotlund L. Moore, *The Galaxy of Aocicinori* (Houston: Commercial Letter Service, 1959). You can read more about it at https://imgur.com/gallery/QldM5.

17. Karen Masters, "Curious? What If the Earth were a Cube?," *Beautiful Stars* (blog), August 18, 2011, http://thebeautifulstars.blogspot.com/2011/08/curious-what-if-earth-were-cube.html.

18. Anders Sandberg, "Torus-Earth," *Andart* (blog), February 4, 2014, http://www.aleph.se/andart/archives/2014/02/torusearth.html.

19. For a detailed exploration of its physics, see Greg Egan, "The World of Dichronauts," Dichronauts, December 11, 2016, http://www.gregegan.net/DICHRONAUTS/01/World.html.

20. For more detail on this, see Greg Egan, "The Perils of Toppling," Dichronauts, September 11, 2017, http://www.gregegan.net/DICHRONAUTS/03/Toppling.html.

21. For an informal but informative discussion of the physics in Inverted World, see "Inverted World, by Christopher Priest," *Em and Emm Expound on Exposition* (blog), March 3, 2011, http://em-and-emm.blogspot.com/2011/03/inverted-world-by-christopher-priest.html.

22. "Literature / Mission of Gravity," *TVTropes*, accessed April 8, 2024, https://tvtropes.org/pmwiki/pmwiki.php/Literature/MissionOfGravity.

23. The image was taken at 5:26 UT (12:26 a.m. EST) on January 1, 2019, when the spacecraft was 4,109 miles (6,628 kilometers) from Ultima Thule (now called Arrokoth) and 4.1 billion miles (6.6 billion kilometers) from earth. For more, see Nola Taylor Tillman, "2014 MU69: Arrokoth, the Most Distant Object Ever Explored," *Space*, September 20, 2022, https://www.space.com/32049-kbo-2014-mu69.html.

24. See, for example, Nadia Drake, "What Is the Multiverse—and Is There Any Evidence It Really Exists?," *National Geographic*, May 4, 2022, https://www.nationalgeographic.com/science/article/what-is-the-multiverse.

Chapter 6: The United Micro Kingdoms (UMK), A Travelers' Tale

1. Originally presented as part of a Graduate Institute for Design, Ethnography, and Social Thought seminar at The New School in New York City on March 4, 2016, based on the UMK project completed in 2013 (pre-Brexit). Some minor modifications have been made to the original unpublished text. The text was written from the perspective of designers interested in material culture and the "stuff" around us—how we interact with it, and its entanglement with alternative worldviews and people's behaviors. Extracts were also published as "United Micro Kingdoms," in *Solution 275-294: Communists Anonymous*, ed. Ingo Niermann and Joshua Simon (Berlin: Sternberg Press, 2018), 187-203.

2. For more on this, see digicar conceptual designer Matthieu Cherubini's work at http://research.mchrbn.net.

3. With thanks to Lana Porter.

Chapter 7: Once Possible, Now Impossible: A Partial Inventory of National Dreams Made Physical

1. Sebastian Jordana, "The Berg: The Biggest Artificial Mountain in the World," *ArchDaily*, November 12, 2009, https://www.archdaily.com/40755/the-berg-the-biggest-artificial-mountain-in-the-world.

2. Iain Treloar, "A Dutchman and His (Man-made) Mountain," *CyclingTips*, May 15, 2020, https://cyclingtips.com/2020/05/a-dutchman-and-his-man-made-mountain.

3. Kaushik Patowary, "The Slag Heaps of Loos-en-Gohelle," *Amusing Planet*, August 5, 2015, https://www.amusingplanet.com/2015/08/the-slag-heaps-of-loos-en-gohelle.html.

4. Lawrence J. Hogan, *Man-Made Mountain* (Sydney: Charter Books, 1979), 68-74.

5. Martin Rees, "Martin Rees on *On the Future*," Princeton University Press, October 1, 2018, https://press.princeton.edu/ideas/martin-rees-on-on-the-future.

6. Quoted in Darren Oldridge, *Strange Histories: The Trials of the Pig, the Walking Dead, and Other Matters of Fact from the Medieval and Renaissance Worlds* (Abingdon, UK: Routledge, 2004), 175.

7. Alec Nevala-Lee, "The Dreamlife of Engineers," *Nevalalee* (blog), March 29, 2018, https://nevalalee.wordpress.com/2018/03/29/the-dreamlife-of-engineers.

8. Elizabeth Colbert, "Can Carbon-Dioxide Removal Save the World?," *New Yorker*, November 13, 2017, https://www.newyorker.com/magazine/2017/11/20/can-carbon-dioxide-removal-save-the-world.

9. The figures used below, gleaned from various online sources, are approximations due to ambiguities over precise start and end dates, imprecise budget translations from one era to another, currency conversions, and other factual uncertainties.

10. The referendum result was announced on Friday, June 24, 2016, at 07:20 BST, and the United Kingdom finally withdrew from the European Union at 23:00 GMT on January 31, 2020.

11. All GDP and GDP per capita figures used in this section are from https://www.worldometers.info/gdp/gdp-by-country.

12. "About Us," Air Force Nuclear Weapons Center, accessed July 29, 2024.

Chapter 8: A Nonstandard, Incomplete Glossary

1. Darko Suvin, "On the Poetics of the Science Fiction Genre," *College English* 34, no. 3 (December 1972): 372-382, http://www.jstor.org/stable/375141.

2. Moon Kyungwon and Jeon Joonho, "The Ways of Folding Space & Flying," Moonandjeon.com, accessed July 25, 2024, https://moonandjeon.com/The-Ways-of-Folding-Space#:~:text=Originating%20from%20Taoist%20practice%2C%20chukjibeop,used%20to%20contract%20physical%20distance.

3. Steven Shaviro, *Extreme Fabulations: Science Fictions for Life* (Cambridge, MA: MIT Press, 2021), 10, Kindle.

4. Yuk Hui and Pieter Lemmens, eds., *Cosmotechnics: For a Renewed Concept of Technology in the Anthropocene* (Abingdon, UK: Routledge, 2021), Kindle.

5. Yuk Hui and Nathan Gardels, "Interview: Singularity vs. Daoist Robots," Research Network for Philosophy and Technology, June 12, 2020, http://philosophyandtechnology.network/3681/interview-singularity-vs-daoist-robots.

6. For more on this, see Jussi Parikka, "Middle East and Other Futurisms: Imaginary Temporalities in Contemporary Art and Visual Culture," *Culture, Theory and Critique* 59, no. 1 (2017): 40-58, https://doi.org/10.1080/14735784.2017.1410439.

7. Levi R. Bryant, *The Democracy of Objects* (Houston: Studium Publishing, 2018), Kindle.

8. H. G. Wells, *The Scientific Romances of H. G. Wells* (London: Victor Gollancz, 1933).

9. Wells, *The Scientific Romances of H. G. Wells*.

10. Bryant, *The Democracy of Objects*.

11. Ursula K. Le Guin, *No Time to Spare: Thinking about What Matters* (New York: Harper, 2017).

12. Quoted in Verlyn Flieger, ed., *Tolkien on Fairy-Stories* (New York: HarperCollins, 2014).

13. Fred Kroon and Alberto Voltolini, "Fiction," Stanford Encyclopedia of Philosophy Archive, Winter 2019 Edition, https://plato.stanford.edu/archives/win2019/entries/fiction.

14. John Law, "What's Wrong with a One-World World?," Heterogeneities, September 19, 2011, http://www.heterogeneities.net/publications/Law2011WhatsWrongWithAOneWorldWorld.pdf.

15. For more artifacts from the story, see "Artifacts Left by Visitors in the Zones," accessed June 23, 2023, http://pid.bungie.org/Roadside_Picnic_Artifacts.html.

16. Shaviro, *Extreme Fabulations*, 18-19.

17. Timothy Morton, "Introducing the idea of 'hyperobjects': A New Way of Understanding Climate Change and Other Phenomena," *High Country News*, January 19, 2005, https://www.hcn.org/issues/47-1/introducing-the-idea-of-hyperobjects.

18. Delphi Carstens interviews Nick Land, "Hyperstition: An Introduction," 0(rphan)d(rift>) Archive, 2009, https://www.orphandriftarchive.com/articles/hyperstition-an-introduction.

19. See Paul Guyer and Rolf-Peter Horstmann, "Idealism," Stanford Encyclopedia of Philosophy Archive, Spring 2023 Edition, https://plato.stanford.edu/archives/spr2023/entries/idealism.

20. Francesco Berto and Mark Jago, "Impossible Worlds," Stanford Encyclopedia of Philosophy Archive, Summer 2023 Edition, https://plato.stanford.edu/archives/sum2023/entries/impossible-worlds.

21. Berto and Jago, "Impossible Worlds."

22. "Irrealism (the Arts)," Wikipedia, last modified May 1, 2023, https://en.wikipedia.org/wiki/Irrealism_(the_arts).

23. Ursula K. Le Guin, "Acceptance Speech for the National Book Foundation's Medal for Distinguished Contribution to American Letters" (filmed at the sixty-fifth National Book Awards, New York, November 2014), https://www.youtube.com/watch?t=12&v=Et9Nf-rsALk.

24. Fred Kroon and Alberto Voltolini, "Fictional Entities," Stanford Encyclopedia of Philosophy Archive, Summer 2020 Edition, https://plato.stanford.edu/archives/sum2020/entries/fictional-entities.

25. Johann Marek, "Alexius Meinong," Stanford Encyclopedia of Philosophy Archive, Fall 2022 Edition, https://plato.stanford.edu/archives/fall2022/entries/meinong.

26. David K. Lewis, *On the Plurality of Worlds* (London: Wiley-Blackwell, 2001).

27. See, for example, Simon O'Sullivan, "Mythopoesis or Fiction as Mode of Existence: Three Case Studies from Contemporary Art," *Visual Culture in Britain* 18, no. 2 (2017): 292-311, doi:10.1080/14714787.2017.1355746.

28. Albert Einstein and Max Born, "Letter from Einstein to Max Born, 3 March 1947," in *The Born-Einstein Letters; Correspondence between Albert Einstein and Max and Hedwig Born from 1916 to 1955* (New York: Walker, 1971), 157-158.

29. "Novum," Wikipedia, last modified January 26, 2022, https://en.wikipedia.org/wiki/Novum.

30. For more on this, see David Ludwig, “Overlapping Ontologies and Indigenous Knowledge: From Integration to Ontological Self-Determination,” *Studies in History and Philosophy of Science Part A 59* (October 2016): 36-45, https://www.sciencedirect.com/science/article/abs/pii/S0039368116300188.

31. Bernardo Kastrup, *Meaning in Absurdity: What Bizarre Phenomena Can Tell Us about the Nature of Reality* (Winchester, UK: John Hunt Publishing, 2012), 20-21, Kindle.

32. As Amiria Henare, Martin Holbraad, and Sari Wastell write, “We want to propose a methodology where the ‘things’ themselves may dictate a plurality of ontologies.” Amiria Henare, Martin Holbraad, and Sari Wastell, eds., *Thinking through Things: Theorizing Artefacts Ethnographically* (Abingdon, UK: Routledge, 2007), 7.

33. Charles L. Harness, *The Rose* (New York: Berkley Medallion, 1969).

34. Ziauddin Sardar, “Welcome to Postnormal Times,” *Futures* 42, no. 5 (June 2010), https://archive.ph/AdhPr#selection-131.0-131.25.

35. For more on this, see Hilary Lawson, “The Poetic Strategy,” IABLIS, 2008, https://www.iablis.de/iablis_t/2008/lawson08.html.

36. For examples of this kind of writing, see *Sci Phi Journal*, https://www.sciphijournal.org.

37. Donna Haraway, *Staying with the Trouble: Making Kin in the Chthulucene* (Durham, NC: Duke University, 2016), 230.

38. Bryant, *The Democracy of Objects*.

39. Philip Ball, *Beyond Weird: Why Everything You Thought You Knew about Quantum Physics Is Different* (Chicago: University of Chicago Press, 2018), 162, Kindle.

40. Karl Ove Knausgaard, *The Morning Star*, trans. Martin Aitken (London: Penguin Books, 2022), 615, Kindle.

41. James Elkins, *Six Stories from the End of Representation: Images in Painting, Photography, Astronomy, Microscopy, Particle Physics, and Quantum Mechanics, 1980-2000* (Redwood City, CA: Stanford University Press, 2008), 217.

42. Ziauddin Sardar, *The Postnormal Times Reader* (Herndon, VA: International Institute of Islamic Thought, 2019), 136, Kindle.

43. Mark Fisher, *The Weird and the Eerie* (London: Repeater, 2016), 15, Kindle.

44. Christopher Menzel, “Possible Worlds,” Stanford Encyclopedia of Philosophy Archive, Summer 2023 Edition, https://plato.stanford.edu/archives/sum2023/entries/possible-worlds.

45. Michael P. A. Murphy, *Quantum Social Theory for Critical International Relations Theorists: Quantizing Critique* (London: Palgrave Macmillan, 2021).

Chapter 9: C/D: By Way of a Conclusion

1. See Ziauddin Sardar, “Welcome to Postnormal Times,” *Futures* 42, no. 5 (June 2010), https://archive.ph/AdhPr#selection-131.0-131.25.

2. See Yuk Hui and Pieter Lemmens, eds., *Cosmotechnics: For a Renewed Concept of Technology in the Anthropocene* (Abingdon, UK: Routledge, 2021), Kindle.

Selected Bibliography

Agron, Melinda, Timon Covelli, Alexis Kandel, and David Langdon, eds. *Perspecta 54: Atopia*. Cambridge, MA: MIT Press, 2022.

Ait-Touati, Frederique, et al. *Terra Forma: A Book of Speculative Maps*. Translated by Amanda DeMarco. Cambridge, MA: MIT Press, 2022.

Anonymous. *The Classic of Mountains and Seas*. Translated by Anne Birrell. London: Penguin Classics, 2000.

Avanessian, Armen, and Suhail Malik. "The Speculative Time Complex." In *The Time Complex: Post-Contemporary*, edited by Armen Avanessian and Suhail Malik. Miami: [NAME] Publications, 2016.

Avanessian, Armen, and Andreas Topfer. *Speculative Drawing: 2011-2014*. London: Sternberg Press, 2014.

Baggini, Julian. *How the World Thinks: A Global History of Philosophy*. London: Granta Books, 2019.

Balakrishnan, Gopal, Ray Brassier, Ted Chiang, Jayce Clayton, Samuel Delany, Silvia Federici, Rivka Galchen, et al. *Speculations (The Future Is ______)*. New York: Triple Canopy, 2015.

Banham, Reyner. *Megastructure: Urban Futures of the Recent Past*. New York: Monacelli Press, 2020.

Barjavel, Rene. *Ashes, Ashes*. N.p.: Modern Literary Editions Publishing, 1967.

Becker, Adam. *What Is Real?: The Unfinished Quest for the Meaning of Quantum Physics*. New York: Basic Books, 2018. Kindle.

Beckett, Chris. *Dark Eden*. New York: Broadway Books, 2014. Kindle.

Bellamy, Edward. *Looking Backward 2000-1887*. Oxford: Oxford University Press, 2009. Kindle. First published 1888.

Bello, Mònica, and José-Carlos Mariátegui. *Quantum: In Search of the Invisible*. Barcelona: Provincial Deputation of Barcelona, 2019.

Berardi, Franco "Bifo." *Futurability*. London: Verso, 2017.

Bernhardt, Chris. *Quantum Computing for Everyone*. Cambridge, MA: MIT Press, 2020.

Blank Space. Fairy Tales: *When Architecture Tells a Story, Volume 1*. N.p.: Blank Space Publishing, 2014.

Bleecker, Julian. *It's Time to Imagine Harder*. Los Angeles: Near Future Laboratory, 2023.

Bleecker, Julian, Nick Foster, Fabien Girardin, and Nicolas Nova. *The Manual of Design Fiction*. Los Angeles: Near Future Laboratory, 2022.

Bottici, Chiara, and Benoît Challand, eds. *The Politics of Imagination*. London: Routledge, 2012.

Braidotti, Rosi, and Maria Hlavajova, eds. *Posthuman Glossary*. London: Bloomsbury Academic, 2018.

Bridle, James. *Ways of Being: Animals, Plants, Machines: The Search for a Planetary Intelligence*. New York: Farrar, Straus and Giroux, 2022. Kindle.

Burwell, Jennifer. *Quantum Language and the Migration of Scientific Concepts*. Cambridge, MA: MIT Press, 2018.

Butler, Philip. *Critical Black Futures: Speculative Theories and Explorations*. London: Palgrave Macmillan, 2021.

Byrne-Smith, Dan, ed. *Science Fiction*. Cambridge: MIT Press, 2020.

Campagna, Federico. *Technic and Magic: The Reconstruction of Reality*. London: Bloomsbury, 2018.

Carroll, Sean. *Something Deeply Hidden: Quantum Worlds and the Emergence of Spacetime*. New York: Dutton, 2019. Kindle.

Catapano, Peter, and Simon Critchley, eds. *Question Everything: A Stone Reader*. New York: Liveright, 2022. Kindle.

Cheng, Ian, ed. *Emissaries Guide to Worlding*. Cologne: Koenig Books, 2018.

Chiang, Ted. *Stories of Your Life and Others*. New York: Vintage Press, 2010. Kindle.

Clarke, Susanna. *Piranesi*. London: Bloomsbury Publishing, 2020. Kindle.

Clement, Hal. *A Mission of Gravity*. Utica, NY: Pyramid Books, 1974. First published 1953.

Coccia, Emanuele. *The Life of Plants: A Metaphysics of Mixture*. Cambridge, UK: Polity, 2018.

Collee, Lauren. "Marxist Astronomy: The Milky Way according to Anton Pannekoek." *Public Domain Review*, October 27, 2021. https://publicdomainreview.org/essay/marxist-astronomy-the-milky-way-according-to-anton-pannekoek.

Crouch, Blake. *Dark Matter: A Novel*. New York: Ballantine Books, 2016. Kindle.

Daston, Lorraine, and Katharine Park. *Wonders and the Order of Nature, 1150-1750*. Princeton, NJ: Princeton University Press, 1998.

Dator, Jim. "What Futures Studies Is, and Is Not." In *Jim Dator: A Noticer in Time, Selected Work, 1967-2018*. Berlin: Springer, 2019.

Davidson, Cynthia, ed. *Log 50: Model Behavior*. New York: Anyone Corporation, 2020.

Davies, William, ed. *Economic Science Fictions*. Cambridge, MA: MIT Press, 2019.

De la Cadena, Marisol, and Mario Blaser, eds. *A World of Many Worlds*. Durham, NC: Duke University Press, 2018.

Delany, Samuel R. *Starboard Wine: More Notes on the Language of Science Fiction*. Middletown, CT: Wesleyan University Press, 2012.

Deutsch, David. *The Fabric of Reality: The Science of Parallel Universes–and Its Implications*. London: Penguin Books, 1998. Kindle.

DeWitt, Richard. *Worldviews: An Introduction to the History and Philosophy of Science*. Hoboken, NJ: Wiley-Blackwell, 2018. Kindle.

"Dialetheism." Wikipedia. Last modified January 3, 2023. https://en.wikipedia.org/wiki/Dialetheism.

Dilnot, Clive. "The Gift." *Design Issues* 9, no. 2 (Autumn 1993): 51-63.

Dunne, Anthony. "A Larger Reality." *In Fitness for What Purpose*, edited by Mary V. Mullin and Christopher Frayling. London: Eyewear Publishing, 2018.

Dunne, Anthony, and Fiona Raby. "Design for the Unreal World." In *Studio Time: Future Thinking in Art and Design*, edited by Jan Boelen, Its Huygens, and Heini Lehtinen. London: Black Dog, 2020.

Dunne, Anthony, and Fiona Raby. *Speculative Everything: Design, Fiction and Social Dreaming*. Cambridge, MA: MIT Press, 2013.

Dunne, Anthony, and Fiona Raby. "Treading Lightly in a World of Many Worlds." In *NGV Triennial 2023*. Melbourne: National Gallery of Melbourne, 2023.

Dunne, Anthony, and Fiona Raby. "United Micro Kingdoms." In *Solution 275-294: Communists Anonymous*, edited by Ingo Niermann and Joshua Simon. Berlin: Sternberg Press, 2018.

Egan, Greg. *Dichronauts*. San Francisco: Night Shade Books, 2017. Kindle.

Elborough, Travis. *Atlas of Improbable Places: A Journey to the World's Most Unusual Corners*. Cartography by Alan Horsfield. London: Aurum Press, 2016.

Elmgreen, Michael, and Ingar Dragset. *Tomorrow: Scenes from an Unrealized Film by Elmgreen and Dragset*. London: Victoria and Albert Museum, 2013.

Enfield, N. J. *Language vs. Reality: Why Language Is Good for Lawyers and Bad for Scientists*. Cambridge, MA: MIT Press, 2022. Kindle.

Eshun, Ekow. *In the Black Fantastic*. Cambridge, MA: MIT Press, 2022.

Flusser, Vilem. *Vampyroteuthis Infernalis: A Treatise, with a Report by the Institut Scientifique de Recherche Paranaturaliste*. Translated by Valentine A. Pakis. Minneapolis: University of Minnesota Press, 2012.

Frase, Peter. *Four Futures: Life after Capitalism*. London: Verso Books, 2016.

Friberg, Jonny, ed. *The Face of God-Making Narratives #2*. Göteborg: HDK, 2017.

Gabriel, Markus. *Why the World Does Not Exist*. Cambridge, UK: Polity, 2015.

Gamow, George. *Thirty Years That Shook Physics: The Story of Quantum Theory*. Mineola, NY: Dover Publications, 1966. Kindle.

Garland, Alex, director. *Devs*, 8 episodes, premiered on March 5, 2020. FX on Hulu.

Geuss, Raymond. *Politics and the Imagination*. Princeton, NJ: Princeton University Press, 2009. Kindle.

Gladstone, Brooke. *The Trouble with Reality*. New York: Workman Publishing, 2017.

Godfrey-Smith, Peter. *Other Minds: The Octopus, the Sea, and the Deep Origins of Consciousness*. New York: Farrar, Straus and Giroux, 2017. Kindle.

Golling, Daniel, and Carlos Mínguez Carrasco, eds. *Kiruna Forever*. Stockholm: ArkDes & Arkitektur förlag, 2020.

Graeber, David. "Radical Alterity Is Just Another Way of Saying 'Reality': A Reply to Eduardo Viveiros de Castro." *HAU: Journal of Ethnographic Theory* 43, no. 1 (Autumn 2015): 1-41. https://doi.org/10.14318/hau5.2.003.

Greene, Brian. *The Hidden Reality: Parallel Universes and the Deep Laws of the Cosmos*. New York: Vintage, 2011.

Greene, Maxine. *Releasing the Imagination: Essays on Education, the Arts, and Social Change*. Hoboken, NJ: Jossey-Basse, 2000.

Gribbin, John. *Six Impossible Things*. Cambridge, MA: MIT Press, 2019. Kindle.

Guyer, Paul, and Rolf-Peter Horstmann. "Idealism." Stanford Encyclopedia of Philosophy Archive, Spring 2023 Edition. https://plato.stanford.edu/archives/spr2023/entries/idealism.

Haraway, Donna. *The Cyborg Manifesto: Science, Technology, and Socialist-Feminism in the Late Twentieth Century*. Minneapolis: University of Minnesota Press, 2016. First published 1985.

Harness, Charles L. *The Rose*. New York: Berkley Medallion, 1969.

Harrington, Nicholas. "Roger Penrose versus the World." Project Q, May 5, 2021. https://projectqsydney.com/roger-penrose-versus-the-world.

Hassler-Forest, Dan. *Science Fiction, Fantasy, and Politics: Transmedia World-Building beyond Capitalism*. Lanham, MD: Rowman and Littlefield Publishers, 2016.

Hayes, Brian. *Infrastructure: A Guide to the Industrial Landscape*. New York: W. W. Norton, 2014.

Heuer, Richards J., Jr. *Psychology of Intelligence Analysis*. Langley, VA: Center for the Study of Intelligence, Central Intelligence Agency, 1999.

Hoban, Russell. *Riddley Walker*. London: Gollancz, 2013.

Holbraad, Martin, Morten Axel Pederson, and Eduardo Viveiros de Castro. "The Politics of Ontology: Anthropological Positions." Society for Cultural Anthropology, January 13, 2014. https://culanth.org/fieldsights/462-the-politics-of-ontology-anthropological-positions.

Hui, Yuk. "On the Limit of Artificial Intelligence." *Philosophy Today* 65, no. 2 (Spring 2021). 339-357. doi. 10.5840/philtoday202149392.

"Hunt-Lenox Globe." Wikipedia. Last modified April 19, 2022. https://en.wikipedia.org/wiki/Hunt-Lenox_Globe.

Hyde, Dominic, Filippo Casati, and Zach Weber. "Richard Sylvan [Routley]." Stanford Encyclopedia of Philosophy Archive, Spring 2021 Edition. https://plato.stanford.edu/archives/spr2021/entries/sylvan-routley.

"Impossible Color." Wikipedia. Last modified January 5, 2023. https://en.wikipedia.org/wiki/Impossible_color.

"Interview: Nicholas Harrington on Polarizing Politics and Epistemological Value of Quantum Theory." *Orders beyond Borders* (blog), April 7, 2021. https://ordersbeyondborders.blog.wzb.eu/2020/12/15/interview-nicholas-harrington-on-polarizing-politics-and-epistemological-value-of-quantum-theory.

Jasanoff, Sheila, and Sang-Hyun Kim, eds. *Dreamscapes of Modernity: Sociotechnical Imaginaries and the Fabrication of Power*. Chicago: University of Chicago Press, 2015.

Jeffries, Richard. *After London, or Wild England*. N.p.: Book Jungle, 2012. Kindle. First published 1885.

Kastner, Ruth E. *Understanding Our Unseen Reality: Solving Quantum Riddles*. London: Imperial College Press, 2015. Kindle.

Kavenna, Joanna. *A Field Guide to Reality*. London: Quercus Publishing, 2016.

Kavey, Allison B. *World-Building and the Early Modern Imagination*. London: Palgrave Macmillan, 2010.

Kemnitz, Charles. "Beyond the Zone of Middle Dimensions: A Relativistic Reading of 'The Third Policeman.'" *Irish University Review* 15, no. 1 (Spring 1985): 56-72.

Khan-Magomedov, Selim Omarovich. Georgii Krutikov: *The Flying City and Beyond*. Translated by Christina Lodder. Chicago: University of Chicago Press, 2015.

Klosterman, Chuck. *But What If We're Wrong?: Thinking about the Present as if It Were the Past*. London: Penguin Books, 2016.

Kohn, Eduardo. *How Forests Think: Toward an Anthropology beyond the Human*. Oakland: University of California Press, 2013.

Lake, Crystal B. *Artifacts: How We Think and Write about Found Objects*. Baltimore: Johns Hopkins University Press, 2020.

Latour, Bruno. *Down to Earth: Politics in the New Climatic Regime*. Cambridge, UK: Polity, 2018.

Latour, Bruno. "'We Don't Seem to Live on the Same Planet...'–A Fictional Planetarium." In *Designs for Different Futures*, edited by Kathryn B. Hiesinger and Michelle Millar. Philadelphia: Philadelphia Museum of Art, 2019.

Leighton, Robert B., and Matthew Sands, eds. *The Feynman Lectures on Physics, Volume III*. Boston: Addison-Wesley Publishing, 1964.

Leonard, Craig. *Uncommon Sense: Aesthetics after Marcuse*. Cambridge, UK: Met Press, 2022. Kindle.

Lewis, Flora. "The Quantum Mechanics of Politics." *New York Times*, November 6, 1983. https://www.nytimes.com/1983/11/06/magazine/the-quantum-mechanics-of-politics.html.

Lewis, Peter J. *Quantum Ontology: A Guide to the Metaphysics of Quantum Mechanics*. Oxford: Oxford University Press, 2016.

Ling, L. H. M. *The Dao of World Politics: Towards a Post-Westphalian, Worldist International Relations*. London: Routledge, 2013.

Littler, Richard. *Discovering Scarfolk*. London: Ebury Press, 2014.

Liu, Cixin. *The Three-Body Problem*. Translated by Ken Liu. New York: Tor Books, 2014. Kindle.

London, Jack. *The Iron Heel*. N.p., 2012. Kindle. First published 1908.

Mandeville, John. *The Book of Marvels and Travels*. Oxford: Oxford University Press, 2012.

Manguel, Alberto, and Gianni Guadalupi. *The Dictionary of Imaginary Places*. San Diego, CA: Harcourt, 2000.

Marcuse, Herbert. "On the Future of Art: Art as a Form of Reality." Reel-to-reel collection, Solomon R. Guggenheim Museum Archives, New York, 1969. https://www.guggenheim.org/wp-content/uploads/2018/08/9009675_01_AB_9009700_01-Art-as-a-Form-of-Reality.pdf.

Marek, Johann. "Alexius Meinong." Stanford Encyclopedia of Philosophy Archive, Fall 2022 Edition. https://plato.stanford.edu/archives/fall2022/entries/meinong.

Matthews, Freddy Dewe. *Bouvetøya: A Cultural History of an Isolated Landmass*. Self-published, 2013.

Maudlin, Tim. *Philosophy of Physics: Quantum Theory*. Princeton, NJ: Princeton University Press, 2019.

McCarthy, Tom. "Writing Machines, on Realism and the Real." *London Review of Books* 36, no. 24 (December 18, 2014). https://www.lrb.co.uk/the-paper/v36/n24/tom-mccarthy/writing-machines.

McHale, Brian. *Postmodernist Fiction*. Abingdon, UK: Routledge, 2003.

Mercier, Thomas Clément. "Uses of 'the Pluriverse': Cosmos, Interrupted–or the Others of Humanities." *Ostium* 15, no. 2 (2019). http://ostium.sk/language/sk/uses-of-the-pluriverse-cosmos-interrupted-or-the-others-of-humanities.

"Mesklin." Wikipedia. Last modified April 12, 2022. https://en.wikipedia.org/wiki/Mesklin.

Middleton, Nick. *An Atlas of Countries That Don't Exist: A Compendium of Fifty Unrecognized and Largely Unnoticed States*. London: Macmillan, 2015.

Miéville, China. *The City and the City*. New York: Del Rey, 2009.

Mindrup, Matthew. *The Architectural Model: Histories of the Miniature and the Prototype, the Exemplar and the Muse*. Cambridge, MA: MIT Press, 2019. Kindle.

Mitrović, Ivica, James Auger, Julian Hanna, and Ingi Helgason, eds. *Beyond Speculative Design: Past–Present–Future*. Split, Croatia: SpeculativeEdu, Arts Academy University of Split, 2021.

Mohrhoff, Ulrich. *The World According to Quantum Mechanics: Why the Laws of Physics Make Perfect Sense After All*. Hackensack, NJ: World Scientific Publishing, 2011.

Monocle. *How to Make a Nation: A Monocle Guide*. Berlin: Gestalten, 2016.

Morton, Timothy. *Realist Magic: Objects, Ontology, Causality (New Metaphysics)*. London: Open Humanities Press, 2013.

Morton, Timothy. *Spacecraft (Object Lessons)*. New York: Bloomsbury Academic, 2021. Kindle.

Nagel, Thomas. "What Is It Like to Be a Bat?" *Philosophical Review* 83, no. 4 (October 1974): 435-450. https://doi.org/10.2307/2183914.

"National Intelligence Estimate." Wikipedia, Wikimedia Foundation. Last modified September 7, 2021. https://en.wikipedia.org/wiki/National_Intelligence_Estimate.

Nichols, Shaun, ed. *The Architecture of the Imagination: New Essays on Pretence, Possibility, and Fiction*. Oxford: Clarendon Press, 2006. Kindle.

Nolan, Daniel, "Modal Fictionalism." Stanford Encyclopedia of Philosophy Archive, Winter 2022 Edition. https://plato.stanford.edu/archives/win2022/entries/fictionalism-modal.

Normals. *Anticipating the Future by Projecting Design Fictions and Hyperlinked Storylines onto a Neo-Brutalist Graphic Novel Series*. Self-published, 2013.

O'Brien, Flan. *The Third Policeman*. Funks Grove, IL: Dalkey Archive Press, 2001.

O'Brien, Flan. *The Third Policeman*. Narrated by Jim Norton. Hong Kong: Naxos Audiobooks, 2009. M4A/M4B, 3:37.

"Ontological Turn." *Scholarly Community Encyclopedia*. Last updated October 19, 2022. https://encyclopedia.pub/entry/30068.

Phillips, Rasheedah. *Black Quantum Futurism: Theory and Practice (Volume 1)*. Philadelphia: AfroFuturist Affair, 2015.

Pinkus, Karen. *Fuel: A Speculative Dictionary*. Minneapolis: University of Minnesota Press, 2016.

Pohl, Frederik. *The Coming of the Quantum Cats: A Novel of Alternate Universes*. New York: Random House Publishing, 1995.

Pollen, Annebella. *The Kindred of the Kibbo Kift: Intellectual Barbarians*. London: Donlon Books, 2021.

Priest, Christopher. *Inverted World*. New York: NYRB Classics, 2012. First published 1974.

Risenfelder, Mark. *The Planet Construction Kit*. Chicago: Yonagu Books, 2010.

Rogers, Adam. *Full Spectrum: How the Science of Color Made Us Modern*. Boston: Mariner Books, 2021. Kindle.

Rojo, Alberto G. "The Garden of the Forking Worlds: Borges and Quantum Mechanics." *Oakland Journal 9* (2005): 69-78. http://hdl.handle.net/10323/7649.

Ronen, Ruth. *Possible Worlds in Literary Theory*. Cambridge: Cambridge University Press, 2009. First published 1994.

Rose, Mark, ed. *Twentieth Century Interpretations of Science Fiction: A Collection of Critical Essays (20th Century Views)*. Hoboken, NJ: Prentice Hall, 1976.

Rovelli, Carlo. *The Order of Time*. New York: Riverhead Books, 2018. Kindle.

Rovelli, Carlo. *Seven Brief Lessons on Physics*. New York: Riverhead Books, 2016. Kindle.

Ryan, Marie-Laure. "From Parallel Universes to Possible Worlds: Ontological Pluralism in Physics, Narratology, and Narrative." *Poetics Today* 27, no. 4 (2006): 633-674. https://doi.org/10.1215/03335372-2006-006.

Sardar, Ziauddin, ed. *Rescuing All Our Futures: The Future of Futures Studies*. Westport, CT: Praeger, 1999.

Sarkis, Hashim, Roi Salgueiro Barrio, and Gabriel Kozlowski. *The World as an Architectural Project*. Cambridge, MA: MIT Press, 2020.

Scheerbart, Paul. *The Perpetual Motion Machine: The Story of an Invention*. Translated by Andrew Joron. Cambridge, MA: Wakefield Press, 2011. First published 1910.

Shaviro, Steven. "Defining Speculation: Speculative Fiction, Speculative Philosophy, and Speculative Finance." *Alienocene: Journal of the First Outernational*, December 23, 2019. https://alienocene.files.wordpress.com/2019/12/sts-speculation.pdf.

Shaw, John K., and Theo Reeves-Evison, eds. *Fiction as Method*. Cambridge, MA: MIT Press, 2017.

Smolin, Lee. *Einstein's Unfinished Revolution: The Search for What Lies beyond the Quantum*. London: Penguin Publishing, 2019. Kindle.

Song, Bing. "Applying Ancient Chinese Philosophy to Artificial Intelligence." Noēma, December 2, 2020. https://www.noemamag.com/applying-ancient-chinese-philosophy-to-artificial-intelligence.

Spiller, Neil. *Radical Architectural Drawing*. London: Wiley, 2022.

Stadler, Robert, and Alexis Valliant, eds. *On Things as Ideas*. London: Sternberg Press, 2017.

Stålenhag, Simon. *Things from the Flood*. Los Angeles: Design Studio Press, 2016.

Suvin, Darko. *Metamorphoses of Science Fiction: On the Poetics and History of a Literary Genre*. New Haven, CT: Yale University Press, 1979.

Thorne, Kip. *The Science of Interstellar*. New York: W. W. Norton, 2014.

Tsing, Anna. *The Mushroom at the End of the World*. Princeton, NJ: Princeton University Press, 2015.

Uncertain Commons. *Speculate This!* Durham, NC: Duke University Press, 2013.

Valla, Clement. "Postcards from Google Earth." Accessed January 11, 2023. http://www.postcards-from-google-earth.com.

Viveiros de Castro, Eduardo, and Yuk Hui. "For a Strategic Primitivism: A Dialogue between Eduardo Viveiros de Castro and Yuk Hui." *Philosophy Today* 65, no. 2 (Spring 2021): 391-400. doi:10.5840/philtoday2021412394.

von Uexküll, Jakob. *A Foray into the Worlds of Animals and Humans*. Translated by Joseph D. O'Neil. Minneapolis: University of Minnesota Press, 2010.

Warburton, Nigel. *A Little History of Philosophy (Little Histories)*. New Haven, CT: Yale University Press, 2011. Kindle.

Warburton, Nigel. *Philosophy: The Basics*. Abingdon, UK: Routledge, 2012. Kindle.

Wendt, Alexander. *Quantum Mind and Social Science*. Cambridge: Cambridge University Press, 2015.

Wertheim, Margaret. *Physics on the Fringe: Smoke Rings, Circlons, and Alternative Theories of Everything*. London: Bloomsbury, 2011. Kindle.

Weschler, Lawrence. *Mr. Wilson's Cabinet of Wonder: Pronged Ants, Horned Humans, Mice on Toast, and Other Marvels of Jurassic Technology*. New York: Vintage Books, 2013. Kindle.

"The Wilderness Act of 1964." US Department of Justice. Last updated May 12, 2015. https://www.justice.gov/enrd/wilderness-act-1964.

Wilk, Elvia. *Death by Landscape*. New York: Soft Skull, 2022. Kindle.

Wolf, Mark J. P. *Building Imaginary Worlds: The Theory and History of Subcreation*. Abingdon, UK: Routledge, 2012.

Yelavich, Susan. *Thinking Design through Literature*. Abingdon, UK: Routledge, 2020.

Zohar, Danah. *The Quantum Society: Mind, Physics, and a New Social Vision*. New York: HarperCollins, 1995.